A to Z
of
Scientists in
Weather and Climate

NOTABLE SCIENTISTS

A to Z
of
Scientists in
Weather and Climate

Don Rittner

☑®
Facts On File, Inc.

A TO Z OF SCIENTISTS IN WEATHER AND CLIMATE

Notable Scientists

Copyright © 2003 by Don Rittner

Facts On File, Inc.
132 West 31st Street
New York NY 10001

Library of Congress Cataloging-in-Publication Data

Rittner, Don.
    A to Z of scientists in weather and climate / Don Rittner.
        p. cm.—(Notable scientists)
    Includes bibliographical references and index.
    ISBN 0-8160-4797-9 (hardcover)
    1. Meteorologists—Biography—Dictionaries. 2. Climatologists—Biography—
    Dictionaries. I. Title. II. Series.
    QC858.A2 R57 2003
    551.5'092'2—dc212002152435

Text design by Joan M. Toro
Cover design by Cathy Rincon
Chronology by Dale Williams

Printed in the United States of America

VB FOF 10 9 8 7 6 5 4 3 2 1

This book is printed on acid-free paper.

To
John F. Roach
Peru Central High School
1961–65
A science teacher who made a difference!

CONTENTS

List of Entries

ACKNOWLEDGMENTS

I am indebted to all those who readily supplied me with their life stories and accomplishments and made helpful suggestions for additions to the book. Special thanks go to Duncan Blanchard, Roger Cheng, and John Day, who allowed me to bounce ideas and questions to them and gave me important feedback.

To the staff at the NOAA Central Library, Gavan Tredoux, Charles Bean Moore, Helen Morgan, John Kasso, Alan Richards, the Archives of the Institute for Advanced Study, Marcia Tucker, Rupert Baker, The Royal Society, EOA Scientific Systems, Inc., Mount Washington Observatory, Range Commanders Council Meteorology Group, and NASA's Earth Observatory, for definitions; NOAA Photo Library; to Greta Wagle for her assistance on bibliographical research; Chris Van Buren; and to Frank K. Darmstadt, executive editor at Facts On File, I give all my sincerest thanks.

INTRODUCTION

The novelist Mark Twain once said, "A great, great deal has been said about the weather, but very little has been done about it." This book is about to prove that there has indeed been a great deal of discussion about the weather and that thousands of people worldwide have been doing a great deal about it! Our current knowledge of climate and weather has been built on foundations created by a wide diversity of people from around the world during the last several thousand years.

Although many readers will only think of weather in terms of the nightly weather forecaster and predictions for the next day's conditions, the study of climate and weather is a much more complicated science today and is part of applied physics. The Egyptians and the Babylonians more than 5,000 years ago attempted to connect weather conditions with astronomical events, and the Greeks developed some of the first theories as to why weather happens and how. It was ARISTOTLE (384–322 B.C.E.) who produced the definitive work of his time by amassing the current state of knowledge in his *Meteorologica,* and which became the bible of all meteorology for the next 2,000 years. Unfortunately, the work of Aristotle and others before him was not based on solid scientific experimentation, but rather on proverbs (observations and folklore) and untested theories. Exceptions in the form of Leonardo DA VINCI (1452–1519), protoscientist–engineer–futurist,

were rare, and it was not until Descartes (1596–1650) established the philosophy of scientific inquiry in the 17th century that the vise of Aristotle's work was finally loosened around the collective throat of world science.

Though Descartes and his followers had the right inclinations, they were hampered by the lack of instruments to present them with the empirical data they desperately needed. A notable exception was the work of Archimedes of Syracuse, a brilliant mathematician. Among his contributions to human knowledge was his discovery of the principle of buoyancy, whose relevance to the science of meteorology is of inestimable importance because the vast majority of all precipitation is the result of buoyant forces operative in the atmosphere. The lack of having empirical data changed in the late 16th and throughout the 18th century with the invention of instruments such as the thermometer by GALILEO (1593), the barometer of Evangelista TORRICELLI (1608–1647), and the hygrometer by Jean De Luc (1773), as well as subsequent "laws" developed by Robert Boyle, Blaise Pascal, Edmund Halley, Sir Isaac NEWTON, and others. All laid the groundwork for the real scientific pursuit of weather and climate.

During the 19th century, attention was given to learning about the forces that shape the wind patterns, large and small, and produce clouds and precipitation. With the invention of the telegraph, engineers created communications networks through which large bodies of information could be gathered and processed. Laws were formulated, tested, and modified. Yet, it was during the 20th century that the advent of flight and the use of computers rapidly transformed the study of weather and climate into a hard science. Satellite technology allowed us to "see" and track weather in action.

The study of weather and climate has taken the big picture and broken it down into more discrete units of study, be it the size of raindrops to large convection cells, and yet, by understanding these smaller pieces of the larger puzzle that is our climate, a broader understanding of how our planet works is emerging.

THE SCIENTISTS

This book is a sample of the men and women who have added to the vast database of meteorological knowledge. You will learn how more than 100 people through the last 2,000 years have contributed to this body of knowledge. What you will find remarkable is the diversity of backgrounds of those who have chosen to study the pieces of the puzzle that makes up our weather past, present, and future.

There is no "typical" scientist who has contributed to the study of weather and climate. Men and women from all walks of life, age, socioeconomic backgrounds, even educational attainment have made great strides in our understanding of the weather. I have chosen this selection of biographies to demonstrate exactly that premise. There are literally thousands from which to choose; some readers may disagree with those whom I have chosen or omitted, but all should agree that this sample proves that anyone with a desire, good-headed thinking, and perseverance can contribute to the field.

The stereotype that scientists are eccentric, highbrow, or antisocial white-robed stuck-in-the-lab individuals can be finally put to rest. Throughout these pages, you will learn how a janitor

formulated some of the first ideas about the ice ages or how one contributor, so torn because his religious views could not mingle with his perceptions of aiding an evolutionist, was driven to suicide. You will read how a promising actor became instead the most revered American scientist ever or about another who was killed by a soldier while he was contemplating the answer to his theory. This book even features a high school drop out who succeeded in being the first to seed rain clouds. You will learn that contributions can exist in the form of a single theory on which others expand, or they may be the invention of a single instrument. Moreover, people have made contributions by collecting and synthesizing the work of others. No matter how small or large a contribution they made to weather and climate, each nevertheless helped make the science what it is today. Like a beautiful musical symphony, each contributes an individual note that makes up the whole composition.

THE ENTRIES

I was assisted in my selections by the help of many of the living scientists featured in this book. *A to Z of Scientists in Weather and Climate* includes important weather and climate scientists whose selection is based not only on their scientific accomplishments, but also, more important, on their life histories. My purpose in mind is to demonstrate that you can make major contributions in any field regardless of your socioeconomic background, be it the son or daughter of a farmer or one of nobility. I believe I have succeeded in that endeavor.

Entries are arranged in alphabetical order by surname, with additional information provided as to birth (and death) dates, nationality, and field(s) of specialization. This is followed by an essay ranging from 750 to 2,000 words that presents the entrant's early history, educational background, positions held, prizes and awards, and major contributions to weather and climate studies. This, of course, takes in the entire field of atmospheric science and meteorology; entries also include those who have studied the currents and depths of the oceans as well as the effects of planetary bodies on the Earth. In essence, the book is arranged in this manner:

Entry Head: Name, birth/date dates, nationality, and field of specialization.

Essay: Essays range in length from 750 to 2,000 words. Each contains basic biographical information that includes where the subject was born, parents, educational background, youth activities and interests, positions held, prizes and awards received, and so on. However, the main focus is the entrant's main contribution to weather and climate. Names in small caps within the essays provide easy reference to other people who have biographies in this book.

In addition to the main alphabetical list, you can search for entrants by scientific field, country of birth, country of major scientific activity, and year of birth. These indexes, along with a list of weather-related websites, and a glossary are found in the back of the book. A to Z of Scientists in Weather and Climate features scientists from around the world who represent many countries. Their contributions in weather and climate come from more than two dozen disciplines that range from anthropology to physics. The oldest contributor, ARISTOTLE, was born in 384 B.C.E. while the youngest contributor, Bard Anton ZAJAC, was born in 1972.

A

Abbe, Cleveland
(1838–1916)
American
Astronomer, Meteorologist

The eldest of seven children, Cleveland Abbe was born in New York City on December 3, 1838, to merchant George Waldo Abbe and his wife Charlotte Colgate. Abbe was educated at the New York Free Academy, now City College of New York, and earned a bachelor's degree in 1857 and a master's degree in 1860. He taught mathematics at Trinity Grammar School from 1857–58 and then moved on to Ann Arbor's Michigan State Agricultural College and the University of Michigan, studying astronomy under Professor Franz Brunnow and eventually teaching mathematics there.

Abbe moved to Massachusetts in 1860 and lived in Cambridge until 1864. He was not allowed to participate in the Civil War due to an eye ailment, so he worked for astronomer Benjamin A. Gould (1824–96) who assigned him to computing telegraphic longitudes for the United States Coast Survey (U.S. Coast and Geodetic Survey). Gould, the first American to receive a Ph.D. in astronomy, and Michigan's Brunnow knew each other, both having been one-time heads of the Dudley Observatory in Albany, New York, so it is possible that the job may have been arranged for Abbe between friends.

While in Cambridge, Abbe met and befriended American meteorologist, William Ferrel, at the American Ephemeris and Nautical Almanac. Ferrel later joined the Geodetic Survey, investigating the general theory of tides, and designed the first tide-predicting machine in 1882.

During the years 1865 and 1866, Abbe visited Russia and studied astronomy with Otto Struve and others at the Nicholas Central Observatory in Pulkovo, home of the largest refracting telescope in the world at that time. Struve had recently become director in 1861 after the retirement of his father, Friedrich Georg Wilhelm Struve. When Abbe returned to America in 1866, he tried to establish in New York an observatory modeled on the Pulkovo Observatory. However, the effort was not fruitful and was unable to interest anyone.

In 1868, Abbe had the opportunity to become the director of the Cincinnati, Ohio, Observatory. He expanded the scope of interests at the observatory and proposed that Cincinnati become the American center for collecting, analyzing, and publishing telegraphic weather observations from around the country and begin to forecast daily. With support from the Cincinnati Chamber of Commerce and help from the Western Union Telegraph Company, Abbe began on September 1, 1869, to publish the daily "Weather Bulletin of the Cincinnati Observatory." These telegraphic weather reports, daily weather maps,

Cleveland Abbe. Abbe was the first professional meteorologist in America and published the first telegraphic weather reports, daily weather maps, and forecasts in the country. *(Courtesy AIP Emilio Segrè Visual Archives)*

Commerce, the creation of a national weather service became a reality.

On February 2, 1870, a joint resolution was passed by Congress and signed into law by President U.S. Grant on the 9th:

> to provide for taking meteorological observations at the military stations in the interior of the continent and at other points in the States and Territories . . . and for giving notice on the northern (Great) Lakes and on the seacoast by magnetic telegraph and marine signals, of the approach and force of storms

The act gained little press coverage or notice. The resolution exists online at http://www.lib.noaa.gov/edocs/WeatherService Resolution.html

The immense task was given to the U.S. Army Signal Service Corps, formed a decade earlier, and now under the direction of Gen. A. J. Myer, chief of the signal service. On November 1, 1870, at 7:35 A.M., the first meteorological reports taken in a systematic fashion were telegraphed by observers—made up of army sergeants—from 24 stations to the new Washington, D.C., based weather service, called by Myer, The Division of Telegrams and Reports for the Benefit of Commerce. This agency evolved from the "Weather Bureau," to the "National Weather Service," and now the U.S. National Weather Service.

Besides using the military, Myer turned to volunteers such as Great Lakes professor Lapham, but in 1871, he hired civilian Abbe as "Special Assistant to the Chief Signal Officer." Abbe prepared the tridaily probabilities (his word for forecasts) of storms, and eventually Abbe earned the nickname "Old Probabilities" as a result of the work. Abbe was now the first official U.S. daily forecaster and full-time civilian working for Myer for the next 10 years. His observations and analysis were extremely accurate, and he gained a reputation for being an expert forecaster, although

and forecasts were the first in the country. In 1870, Abbe married Frances Martha Neal. They had three sons.

The use of telegraphy in weather reports had been championed earlier by Joseph HENRY at the Smithsonian Institution, but these were generally not made available to the public. However, efforts were being generated to create a national weather service, thanks to information provided by Professor Increase A. Lapham of Milwaukee, who was providing Great Lakes weather information to Gen. Halbert E. Paine, congressman for Milwaukee. Along with the support of Gen. A. J. Myer, chief of the Army Signal Service, and the encouragement from the New York City Chamber of the new agency was not unwilling to incorporate folklore such as:

A red sun has water in his eye.
When the walls are more than unusually damp, rain is expected.
Hark! I hear the asses bray, We shall have some rain today.
The further the sight, the nearer the rain.
Clear moon, Frost soon.
When deer are in gray coat in October, expect a severe winter.
Much noise made by rats and mice indicates rain.
Anvil-shaped clouds are very likely to be followed by a gale of wind.
If rain falls during an east wind, it will continue a full day.
A light yellow sky at sunset presages wind. A pale yellow sky at sunset presages rain.

Much of the analysis and forecasting was on Abbe's shoulders until the signal corps could properly train observers. General Myer created the first school of meteorology, later abandoned after his death to become Fort Myer. In the early 1880s, more formal courses in meteorological instruction were designed, and Abbe taught many of the classes.

Abbe oversaw the creation and publication of the *Monthly Weather Review* in 1873 and the *Bulletin of International Simultaneous Observatories*, the latter with the cooperation of meteorologists from other countries. The corps also began to distribute forecasts, known as farmers' bulletins, to thousands of rural post offices to display in front of the post-office buildings. Twenty years later, Abbe was the editor of a much larger version of *Monthly Weather Review* that became, under his direction, one of the leading meteorological journals in the world.

Although General Myer was head of the signal corps, basic research was not a top priority for him. Yet, in 1871, when Abbe joined the corps, he initiated research projects such as the collection of lines of leveling, and by the next year Abbe figured out the altitudes of all the signal-service barometers above sea level. Abbe saw the development of a system of cautionary storm-wind signals using flags, instituted in the summer of 1871 for mariners and cities: "By means of a few flags, white, blue, and black, the probable local weather for the next day is indicated in every town and almost every telegraph and telephone station in the country, so that any one may know what to expect and prepare for." He led the research on tornadoes, humidity, atmospheric electricity, use of balloons, and thermometer exposure and created wet-bulb temperature conversion tables.

Abbe was also instrumental in developing a core research program by building a laboratory and center for research, visiting other research centers around the country, and writing complete reports about what he observed. He particularly liked the research on atmospheric electricity being directed by Henry A. Rowland at Johns Hopkins. Rowland was awarded $200 from the signal corps to help in his laboratory. Abbe also formed the Study Room, a small group of scientists at the bureau who were able to tackle individual problems such as frosts, hurricane paths, floods, and other applied problems.

By 1891, Abbe convinced Daniel C. Gilman, president of Johns Hopkins, to begin a study of meteorology and offered his private library of hundreds of books for Gilman's commitment. Under the joint auspices of Johns Hopkins, the U.S. Weather Bureau, and Maryland Agriculture College, the Maryland State Weather Service was created in May 1891. Abbe taught at Johns Hopkins in 1896, giving lectures on climatology. Abbe also helped to promote the establishment of state weather services throughout his career.

Abbe's publications, both astronomical and meteorological, total almost 300 items. He contributed to current periodicals, supplements to annual reports of the chief signal officer, and encyclopedias such as *Encyclopaedia Britannica*, and he wrote reference books such as *Solar Spots and Terrestrial Temperature*; *A Plea for Terrestrial Physics*; *Atmospheric Radiation*; *Treatise on Meteorological Apparatus*; *Preparatory Studies for Deductive Methods in Meteorology*; a *Treatise on Meteorological Apparatus and Methods* (1887), and *Preparatory Studies for Deductive Methods in Storm and Weather Prediction* (1889). Abbe also translated many foreign works.

From 1891 until his retirement in 1916, Abbe spent his time with the U.S. Weather Bureau in the Department of Agriculture where he was interested in the relationship between climate and crops. He was presented the Royal Meteorological Society's Symons Memorial Gold Medal in 1912 for his lifelong contributions in meteorology and the Hartley Medal of the National Academy of Sciences in 1916 for distinguished public service in establishing and organizing the Weather Service of the United States.

Abbe was one of the leading promoters for introducing standard time in America and was chairman of a committee of the American Meteorological Society that urged this reform until it was finally adopted. Standard time based on time zones was instituted first by the North American railroad industry on November 18, 1883. Standard time in time zones was not established by U.S. law until the Act of March 19, 1918, also called the Standard Time Act, nor did it gain immediate acceptance.

Abbe's wife Frances died in 1908. He married Margaret Augusta Percival the following year. Cleveland Abbe died at home in Chevy Chase, Maryland, in 1916, the same year he retired from the weather service. Flags on the main building of the Department of Agriculture and on the Weather Bureau in Washington, D.C., were flown at half-mast on the day of his funeral.

His brother Robert Abbe (1851–1928) was a plastic surgeon and pioneer in the use of catgut sutures. He was a friend of the Curies, pioneers in radiation, and he became one of the first in America to use radium in treating cancer.

⊠ Alberti, Leon Battista
(1404–1472)
Italian
Philosopher, Architect

The son of an exiled Florentine, Alberti was born in Genoa, Italy, on February 18, 1404. He was instructed in mathematics by his father and then studied law at the school of Barsizia at Padua (Padova) and at the University of Bologna.

In 1432, Alberti was appointed a papal secretary by Pope Eugene IV, went to Rome, and wrote biographies of the saints in Latin for the Catholic Church. He later became a canon of the Metropolitan Church of Florence, in 1447, and abbot of San Sovino, or Sant' Eremita, of Pisa.

Alberti wrote several books on art and architecture such as: *Della statua* (On statues), in 1436; *De re aedificatoria* (On the art of building) in 1452, and the 10-volume *Della pittura* (On painting), in 1453. According to historians, he also was a mapmaker who worked with Paolo dal Pozzo Toscanelli, who supplied Christopher Columbus with maps for his first voyage.

An early mathematics influence from his father did not interfere with his being an architect. In 1450, he wrote *Sul piacere di matematica* (On the pleasures of mathematics) in which he described and illustrated his invention, the first mechanical anemometer, which measured wind velocity or speed. His instrument consisted of a weathervanelike object with a swinging plate, a disk placed perpendicular to the wind. The plate had a scale to measure the deflections caused by the wind. Leonardo DA VINCI described a similar one 50 years later, and Robert HOOKE would invent a similar device 300 years later and is often credited as the inventor of the first anemometer. Clearly, Alberti invented the first one.

Alberti also wrote in 1466 the first book on cryptography, and is considered by some as the Father of Western Cryptology. He invented and published the first polyalphabetic cipher, and designed a cipherdisk comprised of two disks (a fixed outer disc containing the plain text and a moveable inner disc with the corresponding cipher text) to simplify the process of enciphering code. According to historians, this class of cipher was not broken until the 1800s. Interestingly, this form of cipher was popularized and used in the 1950s by the chocolate drink Ovaltine and the Captain Midnight Decoder Badge and would be familiar with anyone belonging to the baby-boom generation.

Aside from his math and meteorological contributions, Alberti is best known for his beautiful works of architecture. Attributed to him are masterful works, such as the chapel of the Rucellai in the church of San Pancrazio (1446), the church of San Francesco at Rimini (1450), the facade of the church of Santa Maria Novella (1456), and the churches of San Sebastiano (1460) and Sant' Andrea, (1470), at Mantua.

Alberti died in Rome on April 25, 1472.

⊠ Amontons, Guillaume
(1663–1705)
French
Engineer, Physicist

Although Guillaume Amontons, the son of a lawyer from Normandy, had no formal scientific education, he nevertheless studied geometry and the sciences and made important contributions in physics and meteorology.

Amontons became deaf at an early age while in a Paris Latin school but was not deterred by this handicap. He went on to design many scientific instruments, such as a hygrometer (1687), an improved barometer (1688), an optical telegraph (1688–95), a conical nautical barometer (1695), a thermic motor (1699), and various air and liquid thermometers including a constant-volume air thermometer (1702). As a career however, he worked in government on various public works projects as an engineer.

The first thermometers that were constructed around 1600 and credited to GALILEO were gas thermometers, based on the expansion or contraction of air, raising or lowering a column of water. Fluctuations in air pressure, however, caused inaccuracies in these early thermometers. In 1695, Amontons improved the gas thermometer by using mercury in a closed column. He also experimented with liquids that we use today, such as alcohol for low temperatures, linseed oil for high temperatures, water, and mercury. It was Daniel FAHRENHEIT who later utilized Amontons's studies of thermal expansion of mercury that would lead him to invent the scaled mercury thermometer in 1704.

Amontons was the first to observe one of the so-called gas laws: the volume of a gas is directly proportional to its temperature when pressure is constant. Yet, it was Jacques-Alexandre-César Charles (1746–1823) in 1787 who discovered that it applied to other gases as well, and it was not until 1802 that another scientist, Joseph Louis Gay-Lussac (1778–1850), published the findings. It now carries the name Charles's law or Gay-Lussac's law, rarely Amontons' law.

Amontons proposed a fast communication system, an early form of telegraphy, by means of transmitted light signals from one station to another. Amontons tried out his optical telegraph with the royal family in attendance sometime between 1688 and 1695. However, the invention of the optical telegraph would not be attributed to him but rather to Claude Chappe who developed the system in the 1790s.

Although he published scientific papers, his only book, *Remarques et experiences physiques sur la construction d'une nouvelle clepsydre, sur les barometres, thermometres, & hydrometres* (Physical remarks and experiments on the construction of a news clepsydre, on the barometers, thermometers and hydrometers) was published in Paris in 1695. He dedicated the book to the Academy of Sciences, of which he was a member. Amontons's paper "De la résistance causée dans les machines" (Resistance caused in the machines) was published in 1699 in *Memoires de l'Académie des Sciences* (Reports of the Academy of Sciences), the results of his studies on friction.

It was Amontons who first established the existence of a proportional relationship between friction force and the mutual force between two bodies in contact (the force of friction between two sliding surfaces is proportional to the normal force between them). This "coefficient-of-friction" is the relationship observed when friction force is divided by normal force and is designated with the Greek letter (μ) mu.

He is credited for discovering the two "laws of friction" in 1699—first, that the frictional force is directly proportional to the normal load; second, that the size of the bodies does not affect the friction—although there is evidence that he rediscovered them since LEONARDO DA VINCI had described the process 200 years earlier, or at least understood it. However, Amontons further believed that friction was predominately a result of the work done to lift one surface over the roughness of the other. On the atomic level, the surface of any solid, no matter how polished, has rough spots sticking out. When two surfaces are rubbed together, one surface of roughness has to work its way over the other surface's roughness.

Three hundred years later, in 1999, physicists at the Johns Hopkins University in Baltimore, Maryland, identified the molecular origins of static friction. They found that hydrocarbon molecules that adsorb on any surface exposed to air actually rearrange to lock contacting surfaces together and produce the static friction force that satisfies Amontons's laws. In 1795, Charles-Augustin de Coulomb (1736–1806) added to the second law of friction (force does not depend on the area of contact) that strength due to friction is proportional to compressive force and not always true for large bodies. Coulomb published the work, and this second law of friction is now known as the Amontons-Coulomb law. Recently two mathematicians, Eric Gerde and Michael

Marder of the University of Texas at Austin, have offered a different explanation as to why surfaces lock together and resist sliding. They state that the key is that microscopic impurities such as dust, dirt and stray molecules, and even fingerprints coat the surfaces and prevent smooth motion.

In 1703, Amontons showed that equal reduction of temperature resulted in equal reductions in pressure and that eventually, at a low-enough temperature, the volume and pressure of air would become zero—making him perhaps the first to recognize the concept of absolute zero. This observation would not become widely known until the following century by other scientists such as Joseph Gay-Lussac and Jacques Charles.

Amontons died from gangrene poisoning in Paris on October 11, 1705.

Ångström, Anders Jonas
(1814–1874)
Swedish
Astronomer, Physicist

Born on August 13, 1814, Anders Ångström was educated at the University of Uppsala and made major contributions in spectroscopy and solar physics. Ångström is known as the "Father of Spectroscopy."

Ångström became a lecturer at the University of Uppsala in 1839 and three years later became an observer at Uppsala Observatory remaining until 1858. While at the observatory, in 1853, he was a pioneer in the observation and study of the spectrum of hydrogen. His optical investigations were formulated into a principle of spectrum analysis that demonstrated that a hot gas emits light at the same frequency as it absorbs it when it is cooled.

After leaving the observatory to become professor in physics at Uppsala University (1858–74), he continued his spectral research and, in 1861, identified hydrogen in the atmosphere of the Sun. His main work was in spectroscopy, but he also investigated the conduction of heat and devised a method of determining thermal conductivity in 1863.

In 1867, he studied the spectrum of the aurora borealis and appears to be the first person to do so. He noticed a similarity between auroral displays and various types of electrical discharges, conditions that could be examined and demonstrated in the laboratory. He was the first to recognize that it was an electrical discharge that was responsible for producing the colorful auroras. In 1868, he used a prism to show that auroral light differs from sunlight in that both have different wavelengths. The following year, he produced an improved map of the solar spectrum, replacing the previously used prism with a diffraction grating and introduced the angstrom unit for measuring wavelength that is still used today. His outstanding *Recherches sur le spectre solaire* (Researches of the solar specter) in 1868 presented an atlas of the solar spectrum with measurements of more than 1,000 spectral lines expressed in units of one ten-millionth of a millimeter, the unit which later became known as the angstrom, in his honor. The angstrom is equal to 100 millionth of a meter and is used to specify radiation wavelengths smaller than a micron.

It was not until the 20th century that spectroscopic investigations, the field he fathered, identified that it was charged oxygen atoms that gave the distinctive green colored light to auroras.

Ångström's sons were also accomplished. Knut Ångström studied ways to measure radiation and was a noted designer of scientific instruments, such as a pyrheliometer that took direct measurement of incident solar radiation. It was adopted as the official international standard in 1905. His other son Anders K. Ångström published two early papers on weather forecasting and values: "Probability and Practical Weather Forecasting" (1919) and "On the Effectivity of Weather Warnings" (1922).

Ångström died June 21, 1874, in Uppsala.

⊠ Anisimov, Sergey Vasilyevich
(1950–)
Russian
Physicist

Sergey Anisimov was born on February 20, 1950, in Russia at Orlovo, town of Murom, to Vasily Pavlovich Anisimov, the leader of a sector at Murom's engineering plant, and Serafima Pavlovna Anisimova. He attended secondary school in Murom, graduating in 1967, and then attended and graduated from the All-Union Machine-building Institute, Radio Faculty, in Moscow in 1972 with a diploma with distinction in radio engineering. He attended and completed post-

Sergey Vasilyevich Anisimov. He currently is working on quasi-stationary spatial-temporary aeroelectrical structures relating to the global electric circuit.
(Courtesy of Sergey Vasilyevich Anisimov)

graduate studies in 1978 with the Institute of Physics of the Earth, Russian Academy of Science receiving a diploma in geophysics. He served as an engineer and later researcher for the Institute from 1978 to 1980 and participated on the 26th Soviet Antarctic Expedition in 1980–81.

In 1985, Anisimov received his Ph.D. in physics and mathematics from Moscow's Institute of Physics of the Earth Russian Academy of Science with a thesis entitled "Experimental investigations of atmospheric electric field variations by natural and artificial influence." His research interests deal with the global electric circuit and coupling of geophysical shells in Earth's electrical environment, atmospheric electricity (aeroelectric), Earth's electromagnetic field, computer databases, and longtime geophysical observations, and geophysical experimental techniques.

The global electrical circuit represents the current contour formed by bottom ionosphere and terrestrial surface-conducting layers, with thunderstorm generators as the basic electrical sources, and the areas of a free atmosphere as zones of returnable currents.

Between 1972 and 1978, Anisimov was on the teaching staff of the All-Union Machine-building Institute as a lecturer and chair of Radio Engineering. He published in Russian his first paper, in 1977, entitled "The vertical electric component of geomagnetic pulsation field" in *Natural Electromagnetic Field of the Earth*. This paper fine-tuned the exact composition of geomagnetic pulsation waves, important for applied geophysics studies. He was also principal investigator (1974–76) of a research project on the "Elaboration of dynamic electrostatic fluxmeter for geomagnetic frequency range" at the Institute of Physics of the Earth. His work here is associated with atmospheric electricity, the global electric circuit and geoelectromagnetic environments, and the creation of a database dealing with aeroelectrical and geoelectromagnetic fields.

He has investigated ultra low-frequency pulsations of the atmospheric electric field by exper-imentation and theoretical modeling. He also discovered aeroelectrical structures that are the fundamental cells of the global electric circuit. Other work includes the detecting of the universal power-law spectrum (common power law of change of energy for aeroelectrical pulsation at frequency scale) for the ultra low-frequency pulsations of the atmospheric electric field. Anisimov has explored the electrodynamic properties of fog and proposed the hierarchy of spatial scales for aeroelectrical structures. He also demonstrated the relationship of atmospheric electric fields and the number of geophysical phenomena at higher and middle latitudes.

Other research areas include the investigations of electric and magnetic fields of artificial sources that were carried out for an experimental model of coupling into geophysical shells (Earth's envelopes [hard, gas, and plasma], including lithosphere, atmosphere, ionosphere, and magnetosphere); effects of a solar eclipse in an electric field were also studied. The response of the atmospheric electric field and current to the variation of conductivity was investigated by numeric modeling. The formation of aeroelectrical altitude profiles by numeric modeling for regular convection was developed as well. For research of aeroelectric and geomagnetic fields, highly sensitive equipment (electrostatic flux meter, induction magnetometer, antenna of type "current collector") and the structural–temporal method technique were developed for research in this area.

Anisimov created a geophysical database (http://geobrk.adm.yar.ru:1352) at the Borok Geophysical Observatory, where geomagnetic field observations are presented from data obtained at their Geoelectromagnetic Monitoring Laboratory as well as from ultra low-frequency (ULF) geomagnetic field pulsations and atmosphere electric field observations. All data can be downloaded without charge.

All of these discoveries add to the new fundamental knowledge (aeroelectrical structures, universal power-law spectrum) about Earth's electrical environment. Finding evidence of coupling of geophysical shells into the global electric circuit, being able to provide the description of electrodynamics properties of fog, and the discovery of aeroelectrical structures as long-lived turbulent formation of aeroelectric field and space charge are major contributions to the field.

He has received numerous grants for his studies dealing with local and global electrical currents in the atmosphere, the creation of the midlatitude information system and database for the Borok geophysical observatory, and experimental, theoretical, and computer modeling investigations of aeroelectrical structures of the global electric circuit.

He has been a member of the International Commission on Atmospheric Electricity since 1996. He married Elena Borisovna Kesova in 1972. They have two daughters. Anisimov has close to 100 scientific papers and communications as well as two inventions in the field of aeroelectrical measurements. His major contributions are the discovery of aeroelectrical structures, long-lived turbulent formation of atmospheric electric field and space charge, and the conception of a fair-weather electric field as an aggregation of structures with global, regional, and local spatial scales, combined with universal, local, and magnetic local-time changeability.

Since 1999, he has been the deputy director and head of the GemM Laboratory at the Borok Geophysical Observatory RAS, in Borok, Yaroslavl, Russian Federation. He currently is working on quasi-stationary spatial-temporary aeroelectrical structures relating to the global electric circuit.

⊠ **Anthes, Richard A.**
(1944–   )
American
Meteorologist

Richard Anthes, president of the University Corporation for Atmospheric Research (UCAR)

Richard A. Anthes. Anthes has made many research contributions in the areas of tropical cyclones and mesoscale meteorology. *(Courtesy of Richard A. Anthes)*

since September 1988 is a highly regarded atmospheric scientist, author, educator, and administrator who has contributed considerable research in mesoscale meteorology. Born in 1944 in St. Louis, Missouri, and raised in Waynesboro, Virginia, Anthes knew as a very young child that he wanted to be a meteorologist.

He earned a bachelor of science degree in meteorology at the University of Wisconsin at Madison and pursued his interest in meteorology by working as a student trainee for the U.S. Weather Bureau at the National Oceanic and Atmospheric Administration during the summers of 1962 through 1967. He discovered that an area of particular interest to him was researching some of nature's most devastating and costly weather phenomena: hurricanes and tropical cyclones. Anthes's master's and doctoral theses obtained in 1967 and 1970 respectively from the University of Wisconsin-Madison, reflected this interest.

In 1971, Anthes started to teach and to conduct research at Pennsylvania State University, where he attained a full professorship in 1978. During this period, he also took a year to conduct research and to teach as a visiting research professor at the Naval Postgraduate School in Monterey, California.

He became the director of NCAR's Atmospheric Analysis and Prediction Division at the National Center for Atmospheric Research (NCAR) in 1981. In 1986, Anthes was selected to become the director of NCAR. His leadership ability, administrative talent, drive for excellence, and vision for the future of the atmospheric and related sciences led him to the presidency of the University Corporation for Atmospheric Research (UCAR) in 1988. UCAR is a nonprofit consortium of 66 member universities that award Ph.D.s in atmospheric and related sciences. UCAR manages NCAR in addition to collaborating with many international meteorological institutions through a variety of programs.

Anthes was elected as an American Meteorological Society (AMS) Fellow in 1979. In 1980, he was the winner of the AMS's Clarence L. Meisinger Award as a young, promising atmospheric scientist who had shown outstanding ability in research and modeling of tropical cyclones and mesoscale meteorology. In 1987, he received the American Meteorological Association's Jule G. Charney Award for his sustained contributions in theoretical and modeling studies related to tropical and mesoscale meteorology. In 2000, he was elected an American Geophysical Union Fellow.

He has participated in or chaired more than 30 different national committees (for agencies such as NASA, NOAA, AMS, NSF, the National Research Council, and the National Academy of Sciences), including his chairmanship on the National Weather Service Modernization Committee from 1996–99. He has published more than 90 peer-reviewed articles and books. One book in particular, *Meteorology* (7th edition, 1997), is widely used at colleges and universities as a general introductory book to the field of meteorology for nonmeteorological majors.

Anthes has made many research contributions in the areas of tropical cyclones and mesoscale meteorology. He developed the first successful three-dimensional model of the tropical cyclone and was the father of one of the world's most widely used mesoscale models, the Penn State–NCAR mesoscale model, now in its fifth generation (MM5). In recent years, he has become interested in the radio occultation technique for sounding Earth's atmosphere and was a key player in the successful proof-of-concept GPS/MET experiment.

Archimedes
(287 B.C.E.–212 B.C.E.)
Sicilian/Greek
Mathematician

Archimedes was born circa 287 B.C. in Syracuse, Sicily, an independent 500-year-old Greek city-state. He was the son of an astronomer named Phidias, and he may have been related to Hieron II, the king of Syracuse. Little is known of his childhood, but he probably studied in the framework of Euclid in Alexandria, Egypt.

He is credited with inventing several war machines used to defend Syracuse, as well as compound pulley systems, a planetarium, and a water screw (a pump now known as Archimedes' screw). He is honored as the greatest mathematician of antiquity, the "father of integral calculus," and, along with Isaac NEWTON (1643–1727) and Carl Friedrich Gauss (1777–1855), one of the three greatest mathematicians of all time. Johannes Kepler, Bonaventura Cavalieri, Pierre de Fermat, Gottfried Leibniz, and Newton all built upon the geometry of Archimedes.

His work *On Floating Bodies* lays down the basic principles of hydrostatics, the branch of physics that deals with fluids at rest and under pressure. One of his theorems, referred today as Archimedes' principle, states that "any body immersed in a fluid is buoyed up by a force equal to the weight of the displaced fluid." According to legend, the elated Archimedes emerged from the public baths where the insight came to him and dashed down the main street of Syracuse—stark naked—shouting "Eureka, eureka, eureka" (I have found it!). This is the principle of buoyancy. According to cloud physicist John DAY, "It should be understood as the basic principle that explains the ascent of air that is necessary for the production of the great majority of our clouds and precipitation."

Not all of his writings survive, but notable works that do are: *On Plane Equilibriums* (two books), *Quadrature of the Parabola*, *On the Sphere and the Cylinder* (two books), *On Spirals*, *On Conoids and Spheroids*, *On Floating Bodies* (two books), *Measurement of a Circle*, *The Sandreckoner*, and *The Method*.

He died in 212 B.C., killed by a Roman soldier while Syracuse was under siege by the Roman army.

⊠ **Aristotle**
(384 B.C.E.–322 B.C.E.)
Greek
Philosopher

Aristotle, a Greek philosopher and scientist, has had more influence on the field of meteorology and, for that matter, on most of the sciences than anyone; it lasted for more than 2,000 years.

Born in 384 B.C. in the Ionian colony of Stagira in ancient Macedonia, Aristotle was the son of Nicomachus, a physician at the court of Mayntas II of Macedon, grandfather of Alexander the Great. At 17, he became a student in Plato's Academy in Athens and stayed there for more than 20 years as a student and teacher. In 347, he moved to the princedom of Atarneus in Mysia (northwestern Asia Minor), ruled by Hermias, and who presided over a small circle of Plato followers in the town of Assos. Aristotle befriended Hermias, joined the group, and eventually married Hermias's niece and adopted daughter Pythias.

In about 342 B.C., he moved to Mieza, near the Macedonian capital Pella, to supervise the

Aristotle. The Greek philosopher's views on weather and climate held for almost 2,000 years, until the invention of scientific instruments gave scientists real empirical data. *(Synnberg Photo-gravure Co., Chicago, Philosophical Portrait Series, Open Court Publishing Co., Courtesy AIP Emilio Segrè Visual Archives)*

education of 13-year-old Alexander the Great. Aristotle returned to Athens in 335 to teach, promote research projects, and organize a library in the Lyceum. His school was known as the Peripatetic School. After Alexander's death in 323, Aristotle was prosecuted and had to leave Athens, leaving his school to Theophrastus. He died shortly after at Chalcis in Euboea in 322 B.C.

His writings were immense, and one of his works influenced the field of meteorology for more than 2,000 years. *Meteorologica* (Meteorology), was written in 350 B.C. and comprised four books, although there are doubts about the authenticity of the last one. They deal mainly with atmospheric phenomena, oceans, meteors and comets, and the fields of astronomy, chemistry, and geography.

Aristotle attempted to explain the atmosphere in a philosophical way and discussed all forms of *meteors*, a term then used to explain anything suspended in the atmosphere. Aristotle discussed the philosophical nature of clouds and mist, snow, rain and hail, wind, lightning and thunder, rivers, rainbows, and climatic changes. His ideas included the existence of four elements (earth, wind, fire, and water) and that each were arranged in separate layers but could mingle.

Artistotle's observations in the biological sciences had some validity, but many of his observations and conclusions in weather and climate were wrong. It was not until the 17th century, with the invention of such meteorological instruments as the hygrometer, the thermometer, and the barometer, that his ideas were disproved scientifically. However, he is considered to be the Father of Meteorology for being the first to discuss the subject at length and for the fact that his ideas were the prevalent ones for 2,000 years.

⊠ Arrhenius, Svante August
(1859–1927)
Swedish
Chemist, Physicist

Svante August Arrhenius was born in Vik (or Wijk), near Uppsala, Sweden, on February 19, 1859. He was the second son of Svante Gustav Arrhenius and Carolina Christina (Thunberg). Svante's father was a surveyor and an administrator of his family's estate at Vik. In 1860, a year after Arrhenius was born, his family moved to Uppsala, where his father became a supervisor at the university. The boy was reading by the age of three.

Arrhenius received his early education at the cathedral school in Uppsala, excelling in biology, physics, and mathematics. In 1876, he entered the University of Uppsala and studied physics, chemistry, and mathematics, receiving his B.S. two years later. Although he continued graduate classes in physics at Uppsala for three years, his studies were not completed there. Instead, Arrhenius transferred to the Swedish Academy of Sciences in Stockholm in 1881 to work under Erick Edlund to conduct research in the field of electrical theory.

Arrhenius studied electrical conductivity of dilute solutions by passing electric current through a variety of solutions. His research determined that molecules in some of the substances, split apart, or dissociated from each other, into two or more ions when they were dissolved in a liquid. He found that while each intact molecule was electrically balanced, the split particles carried a small positive or negative electrical charge when dissolved in water. The charged atoms permitted the passage of electricity and the electrical current directed the active components toward the electrodes. His thesis on the theory of ionic dissociation was barely accepted by the University of Uppsala in 1884, the faculty believing that oppositely charged particles could not coexist in solution. He received a grade that prohibited him from being able to teach.

Arrhenius published his theories ("Investigations on the galvanic conductivity of electrolytes") and sent copies of his thesis to a number of leading European scientists. Russian-German chemist Friedrich Wilhelm Ostwald, one of the leading European scientists of the day and one of the principle founders of physical chemistry, was impressed and visited him in Uppsala, offering him a teaching position, which he declined. However, Ostwald's support was enough for Uppsala to give him a lecturing position, which he kept for two years.

The Stockholm Academy of Sciences awarded Arrhenius a traveling scholarship in 1886. As a result, he worked with Ostwald in Riga, with physicists Friedrich Kohlrausch at the University of Würzburg, Ludwig BOLTZMANN at the University of Graz, and with chemist Jacobus van't Hoff at the University of Amsterdam. In 1889, he formulated his rate equation that is used for many chemical transformations and processes, in which the rate is exponentially related to temperature known as the Arrhenius equation.

He returned to Stockholm in 1891 and became a lecturer in physics at Stockholm's *högskola* (high school) and was appointed physics professor in 1895 and rector in 1897. Arrhenius married in 1894 to Sofia Rudbeck and had one son. The marriage lasted a short two years. Arrhenius continued his work on electrolytic dissociation and added the study of osmotic pressure.

In 1896, he made the first quantitative link between changes in carbon-dioxide concentration and climate. He calculated the absorption coefficients of carbon dioxide and water, based on the emission spectrum of the moon, and also calculated the amount of total heat absorption and corresponding temperature change in the atmosphere of various concentrations of carbon dioxide. His prediction of a doubling of carbon dioxide from a temperate rise of 5–6°C is close to modern predictions. He predicted that increasing reliance on fossil fuel combustion to drive the world's increasing industrialization would, in the end, lead to increases in the concentration of CO_2 in the atmosphere, thereby giving rise to a warming of the Earth.

In 1900, he published his *Textbook of Theoretical Electrochemistry*. In 1901, he and others confirmed the Scottish physicist James Clerk MAXWELL's hypothesis that cosmic radiation exerts pressure on particles. Arrhenius went on to use this phenomenon in an effort to explain the aurora borealis and the solar corona. He supported the Norwegian physicist Kristian Birkeland's explanation of the origin of auroras that he proposed in 1896. Arrhenius also suggested that radiation pressure could carry spores and other living seeds through space and believed that life on Earth was brought here under those conditions. He likewise believed that spores might have populated many other planets, resulting in life throughout the universe.

In 1902, he received the Davy Medal of the Royal Society and proposed a theory of immunol-

Svante August Arrhenius. In 1896 he made the first quantitative link between changes in carbon dioxide concentration and climate. He predicted that increasing reliance on fossil fuel combustion to drive the world's increasing industrialization would, in the end, lead to increases in the concentration of CO_2 in the atmosphere, thereby giving rise to a warming of the Earth. *(Courtesy AIP Emilio Segrè Visual Archives)*

ogy. The following year, he was awarded the Nobel Prize for chemistry for his work that originally was deemed improbable by his Uppsala professors. He also published his *Textbook of Cosmic Physics*.

He became director of the Nobel Institute of Physical Chemistry in Stockholm in 1905 (a post he held until a few months before his death); he married Maria Johansson, with whom he later had one son and two daughters. The following year, he had time to publish three books *Theories of Chemistry*, *Immunochemistry*, and *Worlds in the Making*.

Arrhenius was elected a Foreign Member of the Royal Society in 1911, the same year he received the Willard Gibbs Medal of the American Chemical Society. Three years later, he was awarded the Faraday Medal of the British Chemical Society. He was also a member of the Swedish Academy of Sciences and the German Chemical Society.

During the latter part of his life, his interests included the chemistry of living matter and astrophysics, especially the origins and fate of stars and planets. He continued to write books such as *Smallpox and Its Combating* (1913), *Destiny of the Stars* (1915), *Quantitative Laws in Biological Chemistry* (1915), and *Chemistry and Modern Life* (1919) and received honorary degrees from the universities of Birmingham, Edinburgh, Heidelberg, and Leipzig and from Oxford and Cambridge universities.

He died in Stockholm on October 2, 1927, after a brief illness and is buried at Uppsala.

⊠ **Atlas, David**
(1924–)
American
Meteorologist, Inventor

David Atlas was born on May 25, 1924, in Brooklyn, New York, one of three children born to Rose (Jaffee) Atlas and Isadore Atlas, immigrants from Russia and Poland. Atlas spent the 1930s attending elementary and secondary school in Brooklyn. From age 13 to 17, he played the accordion, hoping to become a virtuoso, but decided not to pursue it. Music may have lost an accordion player, but science gained a major contributor.

Atlas commuted to the highly respected Boys High School in Brooklyn. For several months before college, he worked in the main New York Public Library for $11 a week in the map room and later with their collection on patents. Later in life, his inventive mind would produce more than 20 patents for his own inventions.

He began classes at the City College of New York (CCNY) in September 1941 with the hopes of specializing in electrical engineering, but with

David Atlas. Atlas continues his research on the physics and dynamics of convective storms. *(Courtesy of David Atlas)*

the outbreak of World War II, he accelerated his education, completing an entire year's program of calculus during the summer of 1942. At the age of 18, he obtained a job on the production line making aircraft radios at Western Electric in Kearny, New Jersey.

In the fall of 1942, his second application for premeteorology training for the Army Air Corps was approved, and he spent six months as an enlisted man and nine months as an aviation cadet studying meteorology. He was assigned as a forecaster at a flying school at Courtland, Alabama, where he remained for only three weeks before being reassigned to the Weather Instrument Training School at Seagirt, New Jersey. There, he studied basic electronics for two months before proceeding to Harvard/MIT for radar school. He would spend the rest of his life fine-tuning the application of radar technology for meteorology.

Upon graduation from radar school in April 1945, Atlas was assigned to the newly formed All Weather Flying Division (AWFD) at Wright Field, Dayton, Ohio, to do research and development on weather radar for flight safety. He received his bachelor's of science degree from CCNY in February 1946 and was discharged from the army in October that year.

After a period at Ohio State University, he returned to graduate school at MIT in 1948. There, he received an M.S. in 1951 and D.Sc. in 1955, both in meteorology, while working at the Air Force Cambridge Research Laboratories. After World War II, he became chief of the Weather Radar Branch at the Air Force Cambridge Research Laboratories. Then, in 1966, he became professor of meteorology at the University of Chicago's department of geophysical sciences and also the director of the Laboratory for Atmospheric Probing, the joint laboratory of University of Chicago and Illinois Institute of Technology. He was named director of the Atmospheric Technology Division for the National Center for Atmospheric Research (NCAR) in Boulder, Colorado, and from 1974 to 1976, he served as senior scientist and director of the NCAR's National Hail Research Experiment. From 1977 to 1984, he was chief of the NASA/Goddard Space Flight Center's Laboratory for Atmospheric Sciences in Greenbelt, Maryland.

Atlas has been a member, Fellow, chairman, or president of most of the scientific organizations in his field. Notably, he is a Fellow of the American Meteorological Society (president 1974–75), American Geophysical Union, Royal Meteorological Society, American Association for the Advancement of Science, and American Astronautical Society. His honors and awards are too numerous to list, but he has received most of those given by the American Meteorological Society.

Author of more than 260 papers, his 1964 monograph *Advances in Radar Meteorology* as well as six other books that he edited, including the encyclopedic volume *Radar in Meteorology*, became major reference sources for the field.

According to the American Meteorological Society (AMS), his research has included precipitation measurements, Doppler measurements of air motion, clear-air observations using ultrasensitive radar, and observations from space. His research has been relevant to weather and climate and has ranged from the basic to highly applied.

Atlas has been a leader in the field of meteorology for 60 years, and his research in the field of radar meteorology, as one among a handful of pioneers, is second to none. From 1945 to 2002, he researched virtually every aspect of the field of radar meteorology and has made significant contributions to remote sensing. He has 21 patents, which cover a broad spectrum of radar meteorological technology. His inventions of isoecho contour mapping, the first method of measuring winds by Doppler radar; the so-called Velocity Azimuth Display (VAD); and his method of using conventional air traffic-control radars with Doppler capability to detect microbursts and hazardous low-level windshear all have become major contributions in airline safety. The results of some of his inventions are even seen today as daily TV weather displays in the form of color-coded intensities. His *Radar in Meteorology*, a combined history, text, and quasi-encyclopedia, is invaluable to both students and scientists and will guide the field into future decades.

Today, Atlas continues his research on the physics and dynamics of convective storms (responsible for thunderstorms, tornadoes, and hailstorms), as viewed by sensors such as a combination of modern polarimetric and profiler radars, and aircraft traverses as a means of supporting the estimates of their water and heat budgets as deduced from satellite observations.

B

Beeckman, Isaac
(1588–1637)
Dutch
Physicist

Isaac Beeckman was born in Middleburg, Zeeland, in the Netherlands on December 10, 1588. The son of a candlemaker and water conduit installer, his father had to flee London for religious reasons or intolerance to foreigners. Ironically, his father's father had to flee from Brabant to London because of his religion (they were Protestants). Isaac was apprenticed to his father's factory and later pursued the same trade as an independent artisan in Zierikzee, Province of Zeeland.

Beeckman studied philosophy and linguistics at Leiden from 1607 to 1610 and privately at Saumur in 1612. Self-educated in medicine, he took the exams at the University at Caen and received his M.D. in 1618. He studied a number of fields including physics, astronomy, engineering, and meteorology. From 1619 to 1627, he was a conrector (deputy headmaster) at schools in Rotterdam.

Beeckman was one of the earliest to promote the use of mathematics in physics. Although he was not an author of publications in the sciences, he kept a detailed journal, and he influenced several prominent people through personal relationships and in his position as rector for a school in Dordrecht.

He met and befriended René Descartes, and in 1618, Descartes started to study mathematics and mechanics under Beeckman. Descartes, who became one of the most important and influential thinkers in human history and is sometimes called the founder of modern philosophy, tried to downplay Beeckman's influence later on, but it is clear that Beeckman had a major impact on him. Beeckman also had friendships and correspondence and influenced the likes of Pierre Gassendi, professor of philosophy at the University of Aix; Willebrord Snel, professor of Mathematics at Leiden; surveyor Jan Jansz Stampioen; mathematician Marin Mersenne, and others.

UCLA professor Margaret C. Jacob states that Isaac Beeckman must be recognized as the first mechanical philosopher of the Scientific Revolution:

> There were other mechanists before and contemporary with him, but none of them developed a systematic philosophical approach to mechanical problems, one that speculated as to the atomic construction of matter and designated this mechanical philosophy of contact between bodies as the key to all natural forces, to every aspect of reality from watermills to musical sound.

According to the Institute and Museum of History of Science, in Florence, Italy, in 1626,

Beeckman "determined the relation between pressure and volume in a given quantity of air, discovering that pressure grows at a slightly faster rate than the lessening of volume. In his studies on pneumatics, Beeckman refused the explanation of the rise of water in the pump based on the theory of the *horror vacui* (nature's abhorrence of the void), recognizing that it was air-pressure, which caused this phenomenon."

In 1627, he became the rector of the Latin school at Dordrecht. The following year, he set up the first European meteorological station, obtaining data on precipitation, temperature, wind speed, and direction and compiling astronomical observations. Having close ties and friendships to magistrates in the city, they constructed a tower at the school for his meteorological and astronomical work. He made astronomical observations with Belgian astronomer Philippe van Lansberg.

During this time, many scientists were trying to determine how to determine longitude. Spain and the States General of the United Provinces of the Netherlands offered financial rewards to anyone who could do it. GALILEO had attempted it, and in 1636 Beeckman was appointed to the commission (of the Netherlands) to judge Galileo's proposal to determine longitude.

With his friends Beeckman started a private mechanical "society" called *het collegium mechanicum*. Here, members could apply mechanical interests to navigation issues, windmills, or issues affecting Dutch commerce. He even worked on improving the grinding of lenses for telescopes, keeping with his artisan roots. Beeckman died in Dordrecht, Netherlands, on May 19, 1637.

⊠   **Bentley, Wilson Alwyn**
(1865–1931)
American
Amateur Meteorologist,
Photomicrographer

"Snowflake" Bentley was born on February 9, 1865, and raised on a dairy farm outside Jericho,

Vermont, located between Lake Champlain and Mount Mansfield. There, he lived the life of a self-educated New England farmer. He did not attend school until he was 14 but was taught by his former-schoolteacher mother instead.

Living in the rugged-winter hills of Vermont, he developed a passion for studying snowflakes with the use of a small microscope, loaned to him by his mother. It opened a new world of discovery for him. Although Johannes Kepler in 1611 published a short treatise entitled "On the Six-Cornered Snowflake," asking why snow crystals always exhibit a sixfold symmetry, and Robert HOOKE, using a microscope, in 1665 showed their complexity in the first drawings ever, a Vermont farmer was the first to photograph thousands of them to reveal their real beauty. At the age of 19, Bentley developed the technique of taking photographs of snow crystals. He fitted his microscope with a bellows camera and, on January 15, 1885, became the first person to photograph a single snowflake.

In his lifetime, after looking at some 5,000 individual photographs, he discovered that a single hexagonal shape can have many variations. He wrote, "What magic is there in the rule of six that compels the snowflake to conform so rigidly to its laws?" It was Bentley who coined the phrase that no two snowflakes are identical when he explained in a 1922 *Popular Mechanics Magazine* article that "Every snowflake has an infinite beauty which is enhanced by knowledge that the investigator will, in all probability, never find another exactly like it." His pioneering research revealed that different types of storms produced different kinds of snowflakes.

Bentley turned his attention to the study of rain during the summer months. Meteorologists had been focusing on the study of rainfall to measure quantity and rates but not their actual morphology (form or structure). From 1898 to 1904, he made several hundred measurements of the size of raindrops from different storms, using a shallow pan of wheat flour to catch them, and published the results in the *Monthly Weather*

Wilson Alwyn Bentley. On January 15, 1885, he became the first person to photograph a single snowflake. *(Courtesy of NOAA Image Library)*

Review; he also succeeded on October 30, 1898, in photographing their impressions. He demonstrated that the largest raindrops were about 6 mm (one-quarter inch) in diameter, and he discussed different sizes found in different types of storms, promoting the idea that rain could be formed by melting snow, no snow, or a combination of both. Considering that this duality of origin has only been proven correct in 1976, Bentley was ahead of his time.

Interestingly enough, Philipp Lenard, a German physicist, also studied the properties of raindrops and published his results the same year as Bentley, although neither knew each other. Lenard went on a different research path and received the Nobel Prize in physics for his work with cathode rays in 1905. Bentley continued his interest in snowflakes.

Not only did Bentley write several articles in *Monthly Weather Review,* but he also contributed popular articles about snowflakes in *Country Life, National Geographic, Popular Mechanics,* and *The New York Times Magazine.* In fact, he published more than 60 articles between 1898 and 1931. He even gave lectures about his snowflake work as far away as the Buffalo Museum of Science and the Franklin Institute in Philadelphia. In 1924, Bentley was awarded the first research grant, given by the American Meteorological Society, for his work.

Many in the scientific community did not support him or comment publicly on his work, perhaps due to his lack of credentials, although most had purchased slides of his snowflakes. One exception was Professor George H. Perkins of the University of Vermont, who bought some of his photographs and helped Bentley to write his first essay in 1898 entitled "A study of snow crystals," published in *Appleton's Popular Science Monthly*.

Another supporter of Bentley was Dr. William J. Humphreys, chief physicist for the United States Weather Bureau, who recognized the importance of his work. He started a project to collect and preserve the best Bentley photomicrographs and help him get published. In November 1931, McGraw-Hill published Bentley's book *Snow Crystals*. Humphreys wrote an introduction to the book that contained more than 2,000 of his best photomicrographs.

He was given the nickname "Snowflake" Bentley as a result of his photographs. Duncan BLANCHARD, who wrote a biography on Bentley, gave him the title America's First Cloud Physicist. He is also called The Raindrop Man by Dr. Keith C. Heidorn, who maintains an educational website known as The Weather Doctor (http://www.islandnet.com/~see/weather/doctor.htm).

Bentley died at home from pneumonia two days before Christmas in 1931. Several decades later, *Snowflake Bentley*, a children's book based on his life by Jacqueline Briggs Martin, received the 1999 Caldecott Medal. Bentley's original camera is on exhibit at the Old Red Mill in Jericho, Vermont.

⊠   Bjerknes, Jacob Aall Bonnevie
(1897–1975)
Swedish
Meteorologist

Jacob Bjerknes, like his father Vilhelm BJERKNES, was one of the founders of modern meteorology. Jacob Bjerknes was born on November 2, 1897, in Stockholm, Sweden, to Vilhelm Bjerknes, a Norwegian physicist and meteorologist, and Honoria Sophia Bonnevie. He spent his youth in Stockholm, surrounded by his father's academic friends, including his Aunt Kristine Bonnevie, who was Norway's first woman professor in zoology.

The Bjerknes family moved to Christiania (now Oslo), Norway, in 1907 as Vilhelm became chair at the university, but they moved again in 1913 to Leipzig, Germany. Jacob Bjerknes stayed in Christiania to finish college in 1916 and begin science studies at the Norwegian University. Three years later, he went to Leipzig to join his family and assist his father. Along with him went another Norwegian student, Halvor Solberg.

When a German doctoral student, Herbert Petzold, was killed in 1916, Bjerknes took over his research area of convergence lines in the wind field and made the important discoveries that convergence lines may be thousands of kilometers long, that they drift eastward, and that they are connected with clouds and precipitation. This was published in 1919 and was the beginning of his career in theoretical meteorology.

Because of World War I, Vilhelm, Jacob, and Solberg moved to Bergen, Norway, and established the Bergen School of Meteorology in 1917 along with a new geophysics institute at Bergen Museum. Vilhelm Bjerknes was the professor with two assistants, Jacob Bjerknes and Halvor Solberg. It is here that Jacob Bjerknes presented his now famous frontal-cyclone model.

Bjerknes explained, by studying "squall lines," masses of cold air that abruptly force themselves under warmer air, giving birth to fierce short-lived thunderstorms, how cyclones cross the ocean, lose power, and then regroup as intense storm systems. He found that a squall line is followed by a cyclonic weather system where air rotates around a low-pressure center. By 1919, he had developed an integrated model and invented the terms *warm front* and *cold front* to describe these broader weather patterns.

In 1920, he was appointed head of the Weather Forecasting Office for western Norway and began to forecast with his frontal-cyclone model. Although the model was based mainly on ground observations, he verified the vertical structure of the model in 1922 when he went to Zurich, Switzerland, and collected data from mountain-peak observatories in the Alps. He verified the existence of sloping frontal surfaces up to an altitude of 3,000 meters and received a Ph.D. in philosophy from the University of Christiania in 1924 for this discovery. In 1926, he became a support meteorologist for Norwegian Roald Amundsen's polar dirigible (the *Norge*) flight that sailed over the North Pole. Two years later, he married the daughter of a well-known Bergen ophthalmologist.

In 1928, Bjerknes first described the waves in the upper-tropospheric westerlies, which are connected with the cyclones at low levels. Three years later, he left the weather forecasting office in Bergen and took over a professorship of meteorology especially created for him at the Bergen Museum. In 1939, Bjerknes, with his family, went on a lecture tour to the United States. With the outbreak of World War II and the German invasion of Norway, they ended up staying in America and became U.S. citizens.

He was asked to organize a training school for air-force weather officers at the University of California, and in 1940 he became professor of meteorology at the University of California at Los Angeles (UCLA) and head of meteorology in the department of physics. He brought in Jørgen Holmboe, a Norwegian meteorologist from the Bergen Weather Service. In 1945, Bjerknes became the chairman of a new department of meteorology at UCLA. Under his direction, it became one of the world's leading centers of teaching and research in the atmospheric sciences.

As a result of Bjerknes's and Holmboe's attempt to solve the problem of growing cyclones and their upper waves, one of Bjerknes's students, Jule CHARNEY, produced the first mathematical solution to describe growing waves on a baroclinic current. Baroclinic is a region in which a temperature gradient exists on a constant pressure surface. Charney went on to become one of the leading meteorologists of his day.

With the introduction of computer technology in the 1950s, Bjerknes's cyclone model was put into use when John von NEUMANN and others obtained the first accurate computer-aided weather forecast in 1952 at Princeton using his data. Bjerknes would champion the use of satellites later for imaging weather patterns.

Bjerknes began studies of the Pacific Ocean and the El Niño phenomenon. He found that El Niño is not a local phenomenon but the manifestation of an oscillatory process that affects the atmosphere and the ocean over the entire tropical Pacific. He established a connection between the El Niño phenomenon and the southern oscillation (called ENSO), an irregular pulsation of atmospheric pressure between the Pacific and the Indian Oceans.

Bjerknes was honored with many awards in his life. He was an honorary fellow of the Royal Meteorological Society (1932) and the American Meteorological Society (1966) and was a member of the Royal Norwegian Academy of Sciences, Royal Swedish Academy of Sciences, Danish Academy of Technical Sciences, Academy of Sciences (India), American Academy of Arts and Sciences, and National Academy of Sciences. He was awarded the Royal Meteorological Society Symons Medal (1940); American Geophysical Union Bowie Medal (1945); U.S. Air Force Meritorious Civilian Service Medal (1946); Royal Norwegian Order of St. Olav (1947); Swedish Society of Geography Vega Medal (1958); World Meteorological Organization International Meteorological Organization Prize (1959); American Meteorological Society Carl-Gustaf Rossby Award (1960); Institute of Aerospace Sciences Robert M. Losey Award (1963); and National Medal of Science, (1966). He received an honorary doctor of laws from the University of California in 1967, was president of the Meteorological Association, International Union of Geodesy and Geophysics, 1948–51, and published numerous scientific papers. Although his accomplishments are many, his pioneer work on El Niño is perhaps the most relevant today.

Bjerknes, Vilhelm Friman Koren
(1862–1951)
Norwegian
Mathematician, Meteorologist

Vilhelm Bjerknes was born on March 14, 1862, in Christiania (Oslo), Norway, the son of Carl Bjerknes, a mathematician, and Aletta Koren. Bjerknes was already married in 1883 before he graduated from the University of Christiania (Oslo) in 1888 with an M.A. in math and physics. He moved to Paris in 1889 on a grant and attended the lectures of Jules-Henri Poincaré on the subject of electrodynamics, and was duly impressed with this work.

Later, he moved to Bonn for two years, working as an assistant and collaborator to Heinrich Rudolph HERTZ, a German physicist who pioneered studies on wave theory in electricity. Both men studied electrical resonance that later influenced the development of radio. Bjerknes returned to Norway where he received his Ph.D. in physics in 1892 from the University of Kristiania. (The city, named Christiania when Bjerknes was born, became Kristiania in 1877 and then Oslo in 1925.)

Bjerknes was appointed lecturer (1893) at *högskola* (school of engineering) in Stockholm and then professor of applied mechanics and mathematical physics (1895) at the University of Stockholm. Working with his father, Carl Bjerknes, a mathematics professor at Oslo, they researched the theories of hydrodynamic forces.

In 1897, he developed circulation theorems and a synthesis of hydrodynamics (study of fluid motion and fluid-boundary interaction) and thermodynamics (science of heat and temperature) as they relate to atmospheric and oceanic macroscale motions. He applied the generalizations of the theory of vortices of Thomson and Helmholtz that he discovered to motions in the atmosphere and the ocean. His theory on velocity circulation is known as the Bjerknes theorem.

On November 2, 1897, his son Jacob was born in Stockholm. Jacob, who would be educated

at the University of Kristiania (Oslo) like his father, would later collaborate with him for discovering the mechanism that controls the behavior of midlatitude cyclones (areas of low pressure where winds blow counterclockwise in the Northern Hemisphere and clockwise in the Southern Hemisphere).

In 1904, Vilhelm Bjerknes proposed the idea of numerical weather prediction. In a paper that he published that year, "Weather Forecasting as a Problem in Mechanics and Physics," Bjerknes proposed that with enough information about the current state of the atmosphere, scientists should be able to use mathematics to predict future weather patterns. Mathematician L. F. RICHARDSON tried to do it in 1922 unsuccessfully, and it would take the invention of computers for this idea to take hold; the first successful numerical prediction of weather was made in April 1950, using the ENIAC computer at Maryland's Aberdeen Proving Ground.

The following year, while in the United States, Bjerknes presented his theories of air-mass movements and explained his attempts to apply mathematics to weather forecasting. Impressed with his theories, the Carnegie Institution of Washington began to support financially his research, which lasted until 1941.

In 1906, Columbia University Press published his "Fields of force; supplementary lectures, applications to meteorology; a course of lectures in mathematical physics delivered December 1 to 23, 1905." The following year, he returned to Norway as a professor at the University of Christiania (Oslo) and collaborated with his son Jacob, fellow student Harald Sverdrup, and later Tor Bergeron on the subject of dynamic meteorology.

In 1913, he and Sverdrup moved to Leipzig, Germany, where Bjerknes became professor of geophysics at the University of Leipzig and director of the newly founded Leipzig Geophysical Institute. His son also joined him. However, only four years later, Bjerknes took a position with the Norway museum in Bergen (now part of the

University of Bergen), where he founded the Bergen Geophysical Institute (a weather bureau by 1919).

During World War I, he and his son instituted a network of weather stations throughout Norway. The data collected from these stations eventually led them to their theory of the development of the polar fronts. He published the series *Dynamic Meteorology and Hydrography* (vols. I and II, 1910–11; vol. III, 1951) with the help of his students Johan W. Sandström (on vol. I) and Theodor Hesselberg and Olaf Devik (on vol. II).

They put forward the "Polar front theory of a developing wave cyclone," which states that all weather activity is concentrated in relatively narrow zones called fronts, (analogous to WWW I troop fronts reminiscent of the recent war) and that they form boundaries between warm and cold air masses.

This polar front theory gave meteorologists a way to show how a cyclone evolves from birth, to growth, and finally to death. A cyclone is an area of low pressure around which the winds flow counterclockwise in the Northern Hemisphere and clockwise in the Southern Hemisphere. Meteorologists consider this theory a turning point in the field of meteorology. By the 1940s, fronts were commonly depicted on weather maps. In 1921, he published *On the Dynamics of the Circular Vortex with Applications to the Atmosphere and to Atmospheric Vortex and Wave Motion;* the first modern and comprehensive explanation of the structure and evolution of cyclones.

In 1926, Bjerknes moved again and joined the faculty of the University of Oslo, Norway. He continued work and developed a theory that sunspots are created by an eruption of the ends of magnetic vortices, broken off by the different rotation rates of the Sun's poles and equator (differential rotation). Today, they are known to be concentrations of magnetic flux on the Sun's photosphere and are areas that are cooler than the surrounding photosphere, which is why they appear dark.

In 1929, he published a book on vector analysis, based on part of a larger textbook on theoretical physics that was not completed. He resigned from his professorship at Oslo in 1932, and the following year, he was elected as a Fellow of the Royal Society.

Vilhelm Bjerknes is considered by many to be one of the founders of modern meteorology and weather forecasting and the father of the Bergen school of frontal meteorology (The Bergen School). He has two craters named for him: one on the moon (Crater Bjerknes) and one on Mars (Crater Bjerknes). Bjerknes was nominated for the Nobel Prize in physics 48 times and never won. He died at Oslo on April 9, 1951, at the age of 89.

⊠ Blanchard, Duncan C.
(1924–)
American
Atmospheric Scientist

Duncan Blanchard was born in Winterhaven, Florida, on October 8, 1924, the son of Norman Blanchard and Edna (Perkins) Blanchard. Duncan's father grew vegetables and worked on a small orange plantation before moving to New Lenox, Massachusetts, in 1930, where he accepted a job at the General Electric Company in nearby Pittsfield. As a boy, Duncan loved to commune with nature through the woods near his home and dreamed of becoming a commercial artist like his Uncle Kenneth.

Blanchard began his schooling in a one-room schoolhouse in New Lenox and began Lenox High School but finished his schooling at Pittsfield High in 1942. After delivering papers and working as a theater usher, he obtained his first full-time job in a four-year apprentice course at the Pittsfield General Electric Company. However, with the outbreak of World War II, he became a member of the U.S. Navy and was sent to Harvard and Tufts Universities in an officer-training program (1943–45).

After the war, he went back to Tufts for one year (1946–47), obtaining a bachelor of science degree in general engineering, then to Penn State (1949–51) for an M.S. in physics, and finally to MIT (1956–61) for a Ph.D. in atmospheric science. While at Penn State, he published his first paper in 1950 entitled "The Behavior of Water Drops at Terminal Velocity in Air" that appeared in an issue of *Transactions of the American Geophysical Union*.

Blanchard became part of the famous Project Cirrus, directed by Nobel Prize winner Irving LANGMUIR at the General Electric Research Laboratory (1947–49) in Schenectady, New York, along with a team that included Vincent SCHAEFER and Bernard VONNEGUT. This was the first attempt to seed clouds to create rain.

Blanchard worked at the Woods Hole Oceanographic Institution in Massachusetts from 1951 to 1968 and conducted research on a wide variety of problems such as dealing with issues such as raindrop-size distributions, condensation nuclei, and electrical charges on jet drops from bursting bubbles. He also worked on research dealing with volcanic electricity, the distribution of the marine sea-salt aerosol from the surface of the sea up to cloud base, and the organic material carried by the sea-salt aerosol.

In 1968, he left Woods Hole and became a member of the Atmospheric Research Science Center (ARSC) in Albany, New York, reunited with Schaefer and Vonnegut. He stayed at ARSC for 21 years.

During his years of research, Blanchard made several important discoveries in atmospheric science, including that of an electrical current that is carried by positively charged particles that rise from the sea into the atmosphere. This finding demonstrated that the charge rising from the sea is another important factor in the balance sheet that must be considered with all the other sources of electrical charges in the atmosphere.

He also discovered that when a volcano erupts up through the surface of the sea, the emit-ting cloud that contains droplets of seawater and fragments of lava (ash) carries a positive charge. This new discovery helped shed light on the origin of lightning strokes from volcanoes where the molten lava strikes seawater.

Furthermore, he discovered an extremely high concentration of virus and bacteria on the drops from bubbles that burst at the surface of the sea, lakes, and rivers. This concentration or enrichment is far higher (10 to more than 1,000 times) than that in the waters where the bubbles burst. This had great public-health implications for people who breathe in the aerosol from the bursting bubbles. This aerosol also can damage trees and plants along the shore. Other scientists are studying the mechanism of bacterial and virus enrichment on the droplet aerosol.

Blanchard made many other discoveries but is especially proud of his ability to instill in the public the excitement in understanding science and those who work in the discipline. He has given more than 200 talks to children and adults in this venture and has received numerous invitations to speak at scientific meetings and conventions. *From Raindrops to Volcanoes* (1967), his first book, commissioned by the American Meteorological Society, was an account of his adventures, along with his colleagues, during their first 15 years in science and was translated into six languages. *The Snowflake Man* (1998), his most recent book, was about the life of Wilson BENTLEY, the Vermont farmer who was the first to photograph snowflakes in 1885.

Blanchard married three times, most recently to Julia Nugent in 1984, and has four children. He has been recognized several times for his work throughout his career; for example, he obtained numerous grants from the Office of Naval Research and the National Science Foundations. He was also elected a Fellow of the American Meteorological Society (1980) and the American Association for the Advancement of Science (1982). In 1996, he was selected as the Walter Orr Roberts Lecturer in Interdisciplinary Studies by the American Meteorological Society, and he is listed in *Who's Who in America for Science and Engineering*.

When *Weatherwise* magazine published their 50th anniversary issue in 1998, it contained six articles that the editors considered the most important and interesting from those published within the past 50 years; Blanchard's first article on Wilson Bentley, published in 1970 ("Wilson Bentley, The Snowflake Man"), was selected as one of the six. Besides his two books, Blanchard has published more than 80 technical papers and many nontechnical and educational articles. He retired from ASRC in 1989 and continues to write and exchange email with friends and colleagues.

⊠ **Blodget, Lorin**
(1823–1901)
American
Climatologist

Lorin Blodget was born on May 25, 1823, near Jamestown, New York, and was educated at the local Jamestown Academy and at Hobart (Geneva) College. Blodget's early career was laced with controversy due to personal problems with his superiors, but he finished his career as one of the most astute observers of weather phenomena.

In 1849, Blodget became one of the volunteers for Joseph HENRY's Smithsonian Institution meteorological project, the first network of weather observers. For two years, he contributed data collected from instruments loaned to him by the Smithsonian. With a desire to do more in the field, he asked his local congressmen for help, and at age 28, he became a temporary clerk in November 1851 at the Smithsonian, working for Henry. Blodget was put in charge of analyzing the meteorological data collected by the meteorological project, as well as keeping in personal contact with their meteorological observers. He also

drew maps. Henry was impressed with Blodget's work and promoted him, but Blodget left in 1852 in a dispute with Henry about finances.

Blodget continued working on meteorological projects, and in 1853, he prepared climatological tables for the Seventh U.S. Census, published that year. He contributed to the new field of meteorology by reading several different papers on atmospheric physics at the seventh annual meeting of the American Association for the Advancement of Science, held that year in Cleveland, Ohio. He also appeared that year to be interested in the climatic effects on health issues in a publication "On the Climatic Conditions of the Summer of 1853, Most Directly Affecting Its Sanitary Character," published in the New York Journal of Medicine. Henry rehired Blodget after the AAAS Conference as a clerk, but his tenure was short lived. It appears that Blodget's enthusiasm got in his way of good judgment.

In 1854, Henry fired him from his Smithsonian job for his siding with Assistant Secretary Charles C. Jewett over the creation of a national library at the Smithsonian. Henry locked Blodget out of his office after sending a note to his home telling him that he no longer had a job. It took more than two years for Henry to recoup his meteorological project after letting Blodget go because many of the volunteer observers assumed that the meteorological project was abandoned and many dropped out. Blodget also testified against Henry's meteorological project and requested to the Board of Regents that he be removed from his secretary position. To further embarrass Henry, Blodget published an article in the Patent Office in 1853 that was a compilation of data collected from New York State, the army, and Smithsonian.

For a number of years after the Henry affair, Blodget worked with engineers on the pacific railroad surveys, determining altitudes and gradients and creating meteorological charts. During this period, he further developed the techniques as mapmaker and would forever draw his own maps for his and other publications. His chart "The March of Temperature Changes" from the east base of the Rocky Mountains to Albion Mines, Nova Scotia, in 1854 accompanied reports on exploring for railroad routes from the Mississippi River to the Pacific Ocean. Also in 1854, he published a report on the climatic conditions in the central plains of North America, and, according to University of Oklahoma Professor Bret Wallach, his report was remarkably accurate for the time.

In 1857, he published the first comprehensive American treatment on weather called *Climatology of the United States, and of the Temperate Latitudes of the North American Continent. Embracing a Full Comparison of these with the Climatology of the Temperate Latitudes of Europe and Asia. And Especially in Regard to Agriculture, Sanitary Investigations, and Engineering. With Isothermal and Rain Charts for Each Season, the Extreme Months, and the Year. Including a Summary of the Statistics of Meteorological Observations in the United States, Condensed from Recent Scientific and Official Publications.* It was favorably reviewed in America and in Europe, and considering that much of his data came from his work at the Smithsonian, it appeared to be a coup de grâce for Blodget. To this day, Blodget's *Climatology of the United States* is called "the first work of importance on the climatology of any portion of America, it is so carefully done that all the subsequent works on the subject have confirmed Blodget's major conclusions," by the Mandeville Special Collection Library of the University of California at San Diego.

In 1863, he appears to have left the field of science by moving to the Treasury Department and was placed in charge of preparing several financial and statistical reports. In 1864, "Commercial and Financial Resources of the United States" had 30,000 copies circulated worldwide and boosted the financial reputation of the nation during this period. He spent 12 years as a U.S. appraiser-at-large of customs (1865–77) but

returned for a brief stint as special assistant to the treasury department from 1874 to 1875.

During this time 1858–65, he served as secretary of the Philadelphia Board of Trade, and was the editor of the *North American* (1858–64), a Philadelphia newspaper. In January 1866, he published an article in the *North American Review* titled "Climatology of the United States (Climatic Influences as bearing on succession and reconstruction)" on his theory of what caused succession by the Southern states before the Civil War. Although he ties in climatological differences between the North and the South, the article is a basic call for human rights to former slaves, especially on the issue of voting. He writes:

> But arm this class with the weapon of the free citizen, as silent and more powerful than the billet of a Borgia, and at once all this is changed. He now will be courted, where before he was spurned; interest more potent than philanthropy will now dictate education, where before it demanded debasement; the Pariah will now be treated with outward respect, and the caste harrier be swept away. Without the ballot, four million blacks, increasing more rapidly than the whites, may not only serve as the basis for another aristocracy, fatal to the Union, but also, in the future, bring upon the South the horrors of the war between races so vividly predicted by De Tocqueville.

Blodget's several hundred publications include books, pamphlets, and articles; many of these were reports dealing with finances, revenue, industrial progress, and census, such as "The Census of Industrial Employment, Wages and Social Condition, in Philadelphia, in 1870" (Social Science Association of Philadelphia Papers), "Building Systems for the Great Cities," (*Penn Monthly*, 1877); "The Social Condition of the Industrial Classes of Philadelphia, 1883," and, with William

Moody, "Our labor difficulties microform: the cause, and the way out, including the paper on the displacement of labor by improvements in machinery / by a committee appointed by the American Social Science Association, composed of Lorin Blodget . . . [et al.] in 1878." This was read before the association at their annual meeting in Cincinnati, May 24, 1878, by W. Godwin Moody.

Although his career was tied up with financial and statistical work (he conducted four industrial censuses), he continued to make climatological maps. In 1872, his "Climatological Map of Pennsylvania Showing the Average Temperature, Amount of Rainfall & C. 1872," appeared in the *New Topographical Atlas of the State of Pennsylvania* by Henry F. Walling & O. W. Gray. The following year his "Climatological Map of The State of Maryland" was published by Stedman, Brown & Lyon in Baltimore and appeared in *The New Topological Atlas of the State of Maryland and the District of Columbia*, by S.J. Martenet, H.F. Walling, and O.W. Gray. He also prepared a climatological map of the United States in O. W. Gray's "*Atlas of the United States, with General Maps of the World, 1874.*" He even produced a "*Map to Illustrate Climatology of Canada.*"

⊠ **Boltzmann, Ludwig Eduard**
(1844–1906)
Austrian
Physicist

Ludwig Boltzmann was born on Main Street in the Landstrasse District of Vienna, Austria, on February 20, 1844, the son of a tax man. He attended high school in Linz and was a young naturalist. Tragically, his father died when Boltzmann was only 15.

Awarded a doctorate from the University of Vienna in 1866, Boltzmann's thesis was on the kinetic theory of gases, a subject he spent much of his life researching and writing about. After

Ludwig Boltzmann. Boltzmann became a pioneer in quantum mechanics and founder of statistical mechanics. *(Courtesy AIP Emilio Segrè Visual Archives)*

graduation he became an assistant to his teacher Josef Stefan, who later became known for determining the surface temperature of the sun, using mathematics supplied by Boltzmann, and for his research on radiation. In 1869, Boltzmann was appointed to a chair of theoretical physics at Graz and stayed there for four years.

Boltzmann became a pioneer of quantum mechanics. In 1872, he proposed the "H-theorem" and the Boltzmann equation ($S=klnW$). This latter equation is carved on his gravestone and actually was written by physicist Max Planck. The H-theorem expresses the increase in entropy for an irreversible process and is the mechanical version of the second law of thermodynamics. In 1873, Boltzmann accepted the chair of mathematics at Vienna but was back in Graz after three years as the chair of experimental physics. In 1876, he married Henriette von Aigentler. They had two sons and two daughters.

In 1877, he published "Remarks on Some Problems in the Mechanical Theory of Heat." Here, Boltzmann used statistics to show that entropy could be thought of as a measure of disorder and that the second law of thermodynamics expressed the fact that disorder tends to increase. He founded the subdiscipline of statistical mechanics that deals with the properties of atoms and matter and has the "Boltzmann constant" named for him because of his contributions to the development of this field (the Boltzmann constant has a value of 1.380662×10^{-23} joules per kelvin).

In 1894, Boltzmann moved again, this time back to Vienna as the chair of theoretical physics upon the death of his teacher Josef Stefan. On January 6, 1899, he was elected Fellow of the Royal Academy but moved again the following year to Leipzig to work with Wilhelm Ostwald. Boltzmann unsuccessfully attempted suicide during his time in Leipzig, and some writers attribute this to his disagreements with Ostwald.

In 1902, he returned to Vienna to reclaim his previously held chair of theoretical physics. He also taught philosophy; his lectures became so popular that a big-enough lecture hall could not be found. Boltzmann was one of the first to recognize the importance of Maxwell's electromagnetic theory and formulated the Maxwell-Boltzmann distribution, the average energy of motion of an atom is the same for each direction.

Boltzmann published more than 130 papers during his lifetime, and his work was finally supported by experiments and discoveries in atomic physics that began at the turn of the 20th century.

According to one reviewer of his work, Boltzmann's influence was "central to Max Planck's 1900–1901 papers on blackbody radiation, to Josiah Willard Gibbs's 1902 formulation of statistical mechanics, and to Albert Einstein's 1905 papers on the light quantum and on Brownian motion." Unfortunately, he did not live to see any of it.

He died on October 5, 1906, in Duino (near Trieste), Austria (now Italy), by suicide while on a family vacation. Boltzmann is buried in the Zentralfriedhof (Central Cemetery) in Vienna in Plot: Group 14C, Number 1.

⊠ Bras, Rafael Luis
(1950–)
Puerto Rican/American
Civil Engineer, Hydrologist

Rafael Bras was born in San Juan, Puerto Rico, to Rafael Bras, a civil engineer, and Amalia Antonia (Muniz). After graduating from his high school Colegio San Ignacio in San Juan, he moved to the United States and attended Massachusetts Institute of Technology (MIT) where he obtained a B.S. and M.S. in civil engineering in 1972 and 1974. He married Patricia Brown in 1976; they have two children. He continued at MIT and received his Ph.D. in water resources and hydrology in 1975.

After spending summers as a computer programmer in business applications, Bras began his research career in his area of interest: hydrology. His first paper, coauthored with F. E. Perkins, was "Effects of Urbanization on Catchment Response." Published in 1975 in the *Journal of Hydraulics*, this paper quantified how urbanization increases downstream frequency and peak of floods.

Bras has made several discoveries in the development of real-time flood forecasting and distributed rainfall-runoff models. This work has linked weather predictions to hydrologic forecasting. He also made strides in the effects of quantification of deforestation on local/regional climate that clarify the debate concerning one of the largest anthropogenic (human-made) impacts on climate. He has also developed models of landscape evolution and fluvial (produced by the action of a river or stream) geomorphology that links climate and its landforms.

Bras is a Fellow of the American Geophysical Union (1982), American Meteorological Society (1992), and American Society of Civil believe that he has been successful. He continues his work as Bacardi and Stockholm Water Foundations professor at MIT, where he was elected chair of the faculty. Bras is currently working on the link between landscapes, vegetation, and climate, with increasing interest in the new area of ecohydrology. Ecohydrology bridges the fields of hydrology and ecology and proposes new unifying principles derived from the concept of natural selection.

⊠ **Brook, Marx**
(1920–2002)
American
Physicist

Marx Brook was born in Brooklyn, N.Y., on July 12, 1920; attended the University of New Mexico, receiving his B.S. in physics in 1944; and then attended graduate school at University of California at Los Angeles (UCLA), receiving an M.S. in 1949 and a Ph.D. in physics in 1953.

Brook began to teach physics at New Mexico Tech and to direct the Langmuir Laboratory for Atmospheric Research in Socorro, New Mexico, beginning in 1954. He retired in 1986. During 1960, Brook and others began arrangements to construct a laboratory on the crest of the Magdalena Mountains of what officially became the New Mexico Institute of Mining and Technology.

His primary research has been in atmospheric physics, in the field of lightning and precipitation and other related aspects such as radar. He was concerned with experimental work that leads primarily to explaining the mechanism(s) of charge separation in clouds, and with S.E. Reynolds and M. Gourley, he wrote an important paper on charge separation. Earth and clouds effectively acquire opposite electrical charges, and the air between them serves as an insulator. When the separation of charge aloft is sufficiently large, an ionized path is formed between the cloud and the ground, and a lightning discharge

Rafael Luis Bras. Bras has made several discoveries in the development of real-time flood forecasting and distributed rainfall-runoff models. *(Courtesy of Rafael Luis Bras)*

Engineers (1994). He was also a Guggenheim Fellow. He was awarded the Horton and Macelwane Awards of the American Geophysical Union in 1981 and 1982. He received the Walter L. Huber Civil Engineering Research Prize, ASCE (1993); the Horton Lecturer, American Meteorological Society (1999); and Clarke Prize Laureate for Outstanding Achievement in Water Science and Technology (1998), among others. He is the author of three books on hydrology, has contributed to other books and proceedings, and has published close to 200 scientific papers. He is a member of the U.S. National Academy of Engineering and has an honorary degree from the University of Perugia, Italy.

Bras strives to bring the fields of hydrology and meteorology together in his work, and many occurs, with the charge transferred from the lower part of the cloud (usually negative) to neutralize the induced positive charge on the Earth below. Brook also showed how to reduce the fluctuating radar echo from clouds. He transmitted pulses of broadband noise and averaged the returns within the pulse, rather than average pulse-to-pulse, increasing the scanning rate, so that scanning is limited only by the pulse velocity.

During the summers of 1976, 1977, and 1978, Brook, Paul Krehbiel, and a group of Tech undergraduate students conducted studies at Kennedy Space Center (KSC) as part of their Thunderstorm Research International Program (TRIP). They operated a network of 12 stations at KSC to locate where the lightning discharges were getting their electric charge. They also had access to a network of three Doppler radars as well, and during TRIP, they operated their fast-scanning noise radar that was developed with Brook. It permitted a much greater time resolution of processes within the precipitation-forming regions of clouds.

Brook worked in Japan on the existence of positive groundstrokes. Before his death, he became concerned with elves and sprites, forms of lightning discharge, and how they propagate upward from above cloud to as high as the ionosphere. The ionosphere is a region of the Earth's atmosphere where ionization caused by incoming solar radiation affects the transmission of radio waves. It extends from a height of 50 kilometers (30 miles) to 400 kilometers (240 miles) above the surface. This work was done with Paul Krehbiel and Mark Stanley.

In an age of radars, remote sensors, and digital data-recording devices, Brook has pointed out the usefulness of visual observation and was one of the leaders in using a new vantage point for observing lightning: the space shuttle. Brook was a Fellow of the American Physical Society, the American Geophysical Union, the American Meteorological Society, and the American Association for the Advancement of Science. He died on September 2, 2002, following emergency surgery.

⊠ Buchan, Alexander
(1829–1907)
Scottish
Meteorologist, Oceanographer

Alexander Buchan was born in Kinnesswood (Kinross), Scotland, and was educated at the Free Church College for teachers at Edinburgh and later at Edinburgh University. Although he started out as a teacher and amateur botanist, his interests turned to meteorology.

Buchan was secretary of the Scottish Meteorological Society between 1860 and 1907 and editor of their journal which was first published in 1864. The society merged with the Royal Meteorological Society in 1921. In 1867, Buchan published a textbook, *Handy Book of Meteorology*, that for many years was used and considered to be a standard textbook. He is credited with using the weather map as the basis of weather forecasting because of his work tracing the path of a storm across North America and the Atlantic into northern Europe in 1868. His weather maps were the first to depict the world's average monthly and annual air pressure.

His studies of weather charts resulted in his theory that the climate in the British Isles was subject to annual warm and cold spells that departed from the expected temperature and fell between certain dates each year. He noticed that there were six periods in the year when lower-than-normal temperatures could be expected (in February, April, May, June, August, and November) and three periods where the weather would be higher than normal (June, August, and December). These changes were called Buchan spells. Today, meteorologists believe them to be random occurrences.

He was instrumental in setting up and running the Scottish Meteorological Society's Ben Nevis Observatory, which operated between 1883 and 1904 and provided meteorological data on a continuous hourly basis for 20 years, although it was privately financed. Ben Nevis is the highest mountain in the British Isles (1,344 meters or 4,406 feet) and located in the Grampians in Scotland.

In 1869, his paper "The Mean Pressure of the Atmosphere and the Prevailing Winds Over the Globe, for the Months and for the Year" to the Royal Society of Edinburgh won high praise among the meteorological community. He prepared the 1876 meteorological report of the HMS *Challenger*'s three-and-a-half-year expedition (1872–76) to explore the deep seas and contributed to the oceanographic section. This "Report on Atmospheric Circulation. Report on the scientific results of the exploring voyage of the HMS *Challenger*, 1873–76. Physics and Chemistry, Vol. 2," was published in 1889. This expedition was the first to specifically gather data on a wide range of ocean features that included geology, marine life, ocean current and temperature, and seawater chemistry.

In 1887, he was elected a member of the Meteorological Council and in 1898 was elected a Fellow of the Royal Society. He was the first recipient of the Symons Gold Medal of the Royal Meteorological Society in 1902. He died on May 13, 1907.

Buys Ballot, Christoph Hendrik Diederik
(Christophorus Henricus Didericus)
(1817–1890)
Dutch
Meteorologist

Christoph Hendrik Diederik Buys Ballot was a Dutch meteorologist, born in Kloetinge, the Netherlands, on October 10, 1817. He both studied and taught mineralogy, geology, and mathematics at the University of Utrecht.

In 1854 he founded the Royal Netherlands Meteorological Institute and was its first director. This institute was a result of an outgrowth of a small magnetic observatory lab that he created in

a cellar at Sonnenbergh. On February 1, 1854, he began a small network of observation stations, with four of them outfitted with recording instruments at Utrecht, Groningen, Gelder, and Vlissingen, while the others had regularly scheduled human observations. By 1905, there were 15 stations with instruments and 200 with meteorological observations being recorded.

Yearly meteorological observations were published in the *Netherlands Meteorological Yearbook*, although observations from other countries were also included. The institute also began to publish the first marine meteorological maps in the 1850s, covering the trade winds (1856), North Atlantic winds (1856), general winds (1860), and a series covering windstorms, rain, storms, and fogs (1862), among others.

Buys Ballot was a keen observer. He took two theories by William Ferrel, an American scientist, from the *American Ephemeris and Nautical Almanac*, and Austrian scientist Christian Doppler and expanded on them. In 1855, Buys Ballot conducted a famous experiment to prove the Doppler shift, which relates the frequency of a source to its velocity, relative to an observer, as proposed by Christian Doppler in 1842. Doppler was referring to light sources in stars.

Buys Ballot put a group of musicians on a train with instructions to play and hold a constant note. As the train rushed past him, he was able to hear the Doppler shift, the change in pitch of the music (the note raised as the train neared and fell as it moved past and down the track).

In 1857, he discovered when he stood with his back to the wind that low pressure was on his left and that high pressure was on his right. This effect is only experienced in the Northern Hemisphere. The opposite effect takes place in the Southern Hemisphere. This discovery that the wind blows at right angles to the atmospheric pressure gradient is called the baric wind law, or the Buys Ballot law in his honor. Because Ferrel was first to make the observation and developed a formula for the gradient-wind speed, the law is sometimes called Ferrel's law. This happens because in the Northern Hemisphere, winds circulate counterclockwise around low-pressure areas and clockwise around high-pressure areas. The reverse is true in the Southern Hemisphere.

In 1859, the Dutch government passed a law for gathering and publishing telegraphic reports on the weather, and Buys Ballot and the institute at Utrecht was put in charge of it. Buys Ballot and the institute issued the first regularly issued storm warnings beginning on June 1, 1860. By trying to establish rules for weather forecasting, Buys Ballot may have been the first to connect points of equal pressure on a weather map (isobars), showing zones of low pressure (depressions) and high pressure areas (anticyclones).

He also invented an aeroklinoscope, an instrument that was on shore for sailors to see and use to signal the magnitude and direction of the baric gradient. These gradients cause the acceleration of local intense winds, and sailors could use his Buys Ballot law for staying clear of inclement weather. Buys Ballot died in February 3, 1890, in Utrecht.

Cane, Mark Alan
(1944–)
American
Climatologist, Oceanographer

Born in Brooklyn, New York, on October 20, 1944, Mark Alan Cane is the son of Philip Cane, an electrical engineer who designed traffic signals as well as being a high school teacher and principal, and Ida Edelsberg, a social worker and attendance teacher. He attended Brooklyn schools P.S. 152, Ditmas Junior High School, and Midwood High School, graduating in 1961. As a child, his primary interest was baseball—both to play and to root for the Brooklyn Dodgers. He also liked to read, especially fiction and biographies, and math was his favorite subject in school.

Cane attended Harvard University from 1961 to 1965 and received a bachelor of arts degree in applied math in 1965, and an M.S. in applied math in 1968, the same year he married Barbara Haak. He later attended Massachusetts Institute of Technology and in 1976 received his Ph.D. in meteorology and physical oceanography; the same year, he published his first paper with E.S. Sarachik, "Forced Baroclinic Ocean Motion I: The Equatorial Unbounded Case," which appeared in the *Journal of Marine Research*. It derived some results in linear equatorial-wave theory that are useful in understanding and describing how the low latitude ocean reacts to changing winds.

From 1966 to the 1990s, he served in various capacities that range from program analyst at Goddard Institute for Space Studies to professor in the Department of Earth and Environmental Sciences at Columbia University. During the 1980s, Cane and colleague Stephen Zebiak devised the first numerical model able to simulate El Niño and the Southern Oscillation (ENSO), a pattern of interannual climate variability centered in the tropical Pacific but with global consequences. In 1985, their model was used to make the first physically based forecasts of El Niño. Supported by the Lamont director, they made a public forecast of an imminent El Niño in spring 1986, inaugurating the era of seasonal to interannual climate prediction. For years, the Lamont model has been the primary tool used by many investigators to enhance understanding of ENSO.

Cane has also worked extensively on the impact of El Niño on human activity, especially agriculture. A 1994 *Nature* paper showing the strong effect that El Niño has on the maize crop in Zimbabwe has been influential in prompting decision makers to factor climate variability into their deliberations. The work carried out at Lamont for many years was instrumental in the creation of the International Research Institute for Seasonal to Interannual Climate Prediction.

Mark Alan Cane. Cane is recognized for developing the first physically based model to simulate and explain El Niño, as well as the first physically based predictions of El Niño, as he is for his pioneering work on impacts of climate variations on agriculture and for his contributions to equatorial ocean theory. *(Courtesy of Mark Alan Cane)*

Cane has published more than 170 scientific articles in journals and has received numerous honors, including the Honorary John Harvard Scholarships, National Science Foundation Creativity Award 1984–86, and the Sverdrup Medal of the American Meteorological Society in 1992. He is a Fellow of the American Meteorological Society (1993), the American Association for the Advancement of Science (1995), the American Geophysical Union (1995), and the American Academy of Arts and Sciences (2002).

A speaker, member, and chairperson for many scientific committees, he has been mentor to many masters and doctoral candidates. Cane is recognized for developing the first physically based model to simulate and explain El Niño, as well as the first physically based predictions of El Niño, for his pioneering work on impacts of climate variations on agriculture, and for his contributions to equatorial ocean theory. He is currently the G. Unger Vetlesen Professor of Earth and Climate Sciences at the Lamont-Doherty Earth Observatory of Columbia University in Palisades, New York.

⊠ Celsius, Anders
(1701–1744)
Swedish
Astronomer, Physicist

Anders Celsius was a Swedish astronomer, physicist, and mathematician who introduced the Celsius temperature scale that is used today by scientists in most countries. He was born in Uppsala, Sweden, a city that has produced six Nobel Prize winners. Celsius was born into a family of scientists, all originating from the province of Hälsingland. His father Nils Celsius was a professor of astronomy as was his grandfather Anders Spole; his other grandfather, Magnus Celsius, was a professor of mathematics. Both grandfathers taught at the University in Uppsala. Several of his uncles were also scientists.

It is not surprising that the 29-year-old Anders, who was gifted in mathematics, was appointed professor in astronomy at the university in 1730, taking over for his father who died in 1724. Celsius spent four years touring all of the important European observatories, meeting the scientists of the day, and making observations especially on the phenomena of the aurora borealis, or "northern lights." These visual displays are caused by charged particles, namely electrons and protons, that originate outside the atmosphere that interact with atoms of the upper atmosphere to form the auroras. In 1733, he published a report of 316 observations of the aurora borealis that were made by him and others from 1716 to 1732 and that were some of the first published ever on the subject. Later on, he and his assistant Olof Hiorter were the first to realize that auroras were caused by magnetic effects, based on their observing their effects on compass needles.

After returning to Uppsala, he joined French astronomer Pierre-Louis de Maupertuis's famous 1736 Lapland Expedition to the River Torne Valley in northern Sweden. The expedition's goal was to measure the length of a degree along a meridian, close to the pole, and compare the result with a similar expedition led by Charles-Marie de La Condamine who was sent to Mitad del Mundoto, Peru, near the equator, now part of Ecuador.

The success of the expedition, financed by the French Royal Academy of Sciences, confirmed Isaac NEWTON's opinion that the shape of the Earth is an ellipsoid, flattened at the poles. It was Celsius who had actually suggested the expedition while in Paris to end the debate. The king of France rewarded him with an annual pension of 1,000 livres (one livre was worth a pound of silver). The results of the measurements were published by Maupertuis in his 1738 book *La Figure de la Terre* (The figure of the earth).

The fame Celsius received from the expedition allowed him to convince Swedish authorities to supply money and materials to build an observatory in the center of Uppsala. The Celsius Observatory was completed in 1741 and was equipped with instruments he purchased while travelling. The observatory lasted until 1852 and the building still stands. The current observatory built in 1853 still houses some of Celsius's instruments.

Celsius's important contributions include determining the shape and size of the Earth; gauging the magnitude of the stars in the constellation Aries; publication of a catalog of 300 stars and their magnitudes; observations on eclipses and other astronomical events; and a study revealing that the Nordic countries were slowly rising above the sea level of the Baltic. His most famous contribution falls in the field of meteorology, and the one for which he is remembered most is the creation of the Celsius temperature scale.

In 1742, he presented to the Swedish Academy of Sciences his paper "Observations on Two Persistent Degrees on a Thermometer" in which he presented his observations that all thermometers should be made on a fixed scale of 100 divisions (centigrade), based on two points, 0 degree for boiling water, and 100 degrees for freezing water. He presented his argument on the inaccuracies of existing scales and calibration methods and correctly presented the influence of air pressure on the boiling point of water.

After his death, the scale which he designed was reversed, giving rise to the existing 0° for freezing and 100° for boiling water, instead of the reverse. It is not known if this reversal was done by his student Martin Stromer, by botanist Carolus Linneaus who in 1745 reportedly showed the senate at Uppsala University a thermometer so calibrated, or by Daniel Ekström who manufactured most of the thermometers used by both Celsius and Linneaus. However, Jean Christin from France made a centigrade thermometer with the current calibrations (0° freezing, 100° boiling) a year after Celsius and independent of him and so may therefore equally claim credit for the existing Celsius thermometers.

For years, Celsius thermometers were referred to as centigrade thermometers. However, in 1948, the Ninth General Conference of Weights and Measures ruled that "degrees centigrade" would be referred to as "degrees Celsius," in his honor. The Celsius scale is still used today by most scientists.

Anders Celsius was secretary of the oldest Swedish scientific society, the Royal Society of Sciences in Uppsala between 1725 and 1744, and published much of his work through that organization, including a math book for youth in 1741. He died of tuberculosis in April 25, 1744, in Uppsala.

Chapman, Sydney
(1888–1970)
British
Geophysicist, Mathematician

Sydney Chapman was born on January 29, 1888, in the town of Eccles, Lancashire, England, to the parents of textile cashier Joseph Chapman and Sarah Louisa Gray. His schooling included the higher grade school in Patricroft until age 14, the

Royal Technical Institute in Salford, and then the University of Manchester at age 16. During these early years (1902–05), he studied solar and lunar diurnal magnetic variations.

He received a bachelor of science degree in engineering in 1907, an M.S. in physics in 1908, and a Ph.D. in 1912 from Manchester. He continued his education at Trinity College from 1908 to 1911, where he received a B.A. in Mathematics in 1911 and an M.A. in 1914. While a student at Cambridge, in 1910, he became chief assistant of the Royal Greenwich Observatory at Greenwich. Chapman was in charge of creating a new geomagnetic observatory; this experience shaped his future research. He held this position until 1918. After obtaining his M.A. from Cambridge University in 1914, he became a lecturer at Trinity College.

In 1916 and 1917, he published a complete theory of nonuniform gases and a study of gas mixtures, predicting thermal diffusion, entitled "The Kinetic Theory of Simple and Composite Monatomic Gases; Viscosity, Thermal Conduction, and Diffusion," "On the Kinetic Theory of a Gas. Part II.—A Composite Monatomic Gas: Diffusion, Viscosity, and Thermal Conduction," and "A Note on Thermal Diffusion."

Chapman went on to become a professor of mathematics and natural philosophy at Manchester in 1919 but stayed only five years. In 1922, he married Katherine Nora Steinthal and had four children (three boys, one daughter). In 1924, he accepted a position of professor at the Imperial College of Science and Technology of the University of London and stayed there until 1946 when he accepted the Sedleian professorship of natural philosophy at Queens College at Oxford University. The Royal Society of London elected him a Fellow in 1919. He was the recipient of the Smith and Adams prizes awarded by Cambridge in 1929.

Chapman's research contributed to the knowledge of auroras starting in the early 1920s by explaining their effect on the Earth's magnetism, including the fact that auroral activity

coincided with the periods of greatest sunspot activity and magnetic storms. He also explained that auroras were produced by fast-moving charged particles from the Sun entering the atmosphere and directed toward the poles due to the planet's magnetism. He also made contributions in the composition of the ionosphere.

In 1930, he described a process now known as the Chapman reactions as "Stratospheric reactions in which ozone dissociates into molecular oxygen and atomic oxygen, and the resulting free oxygen atoms recombine with ozone to form molecular oxygen."

Chapman's theory of magnetic storms was formulated in the late 1920s and early 1930s. He and his associate Vincent Ferraro proposed that magnetic storms are caused when plasma clouds ejected from the Sun envelop the Earth ("A New Theory of Magnetic Storms," *Nature*).

The Royal Society of London awarded him its Royal Medal in 1934, an award given for the most important discoveries in the previous year. He helped create the Meteorological Research Committee in Great Britain and was its first chairman from 1941 to 1947. During the 1950s, his research on the composition of the upper atmosphere revealed that volumes of air, like the oceans of the sea, have tides that rise and fall twice a day. These tides, however, are not influenced by the moon but rather by the Sun, although the moon does have some effect on their regular oscillation.

Chapman retired from Oxford in 1953 and became a visiting professor at universities in the United States, Turkey, and Egypt. He presided over the International Geophysical Year (1957–58), heading the committee that was responsible for planning and organizing the year's observations. More than 60 countries coordinated and shared their studies of Earth sciences, under the auspices of the International Council of Scientific Unions. Chapman wrote *The IGY, Year of Discovery*, a popular book that won the Thomas Alva Edison Foundation Award in 1959 as the best science book for youth.

In addition to contributing more than 400 articles to scientific journals on such varied subjects as meteorology, astronomy, physics, and mathematics, Chapman wrote several books including: *The Earth's Magnetism* (1936) and *The Mathematical Theory of Non-uniform Gases; An Account of the Kinetic Theory of Viscosity, Thermal Conduction and Diffusion in Gases*, this last with T. G. Cowling (1940). *Solar Terrestrial Physics* with Syun-Ichi Akasofu was published in 1972.

Chapman delivered a celebrated annual Royal Society Bakerian lecture in 1931 on "Some phenomena in the upper atmosphere," was a Fellow and president of the London Mathematical Society from 1929 to 1931, and received their premier De Morgan Medal in 1944 for his contributions in mathematics. In recognition of his outstanding achievements, Chapman was named a Fellow of the Cambridge Philosophical Society and the American Physical Society.

He was also a member of the Royal Astronomical, Royal Meteorological, and London Physical societies and the National Academy of Sciences of Norway, Sweden, and Finland. In 1951, Chapman was president of the International Union of Geodesy and Geophysics. He received an Honorary Degree of doctor of science from the University of Alaska in 1958. The American Geophysical Union chose Chapman for its William Bowie Medal in 1962. This medal "acknowledges an individual for outstanding contributions to fundamental geophysics and for unselfish cooperation in research."

In recognition of his significant contributions to meteorology, Chapman was appointed by the American Meteorological Society as the first Wexler Memorial Lecturer in 1964. That year Syun-Ichi Akasofu and Chapman revived and expanded Kristian Birkeland's notion of a polar magnetic storm, now named the magnetospheric substorm.

The Royal Society of London awarded him their Copley Medal in 1964. The following year, he was awarded the Royal Meteorological Society's Symons Memorial Gold Medal, and the Smithsonian Institution gave Chapman its Hodgkins Medal, awarded for achievements in atmospheric research.

Chapman is given credit for coining the words *geomagnetism* (the magnetism of the Earth) and *aeronomy* (the study of the upper atmosphere, especially of regions of ionized gas). A crater on the moon is named for him, Crater Chapman. He died at the age of 82 on June 16, 1970, in Boulder, Colorado.

⊠ **Charney, Jule Gregory**
(1917–1981)
American
Meteorologist

Jule Gregory Charney was born in San Francisco, California, on January 1, 1917, to Russian immigrants Ely Charney and Stella Littman, both of whom worked in the garment industry. Charney lived in Los Angeles and attended public schools there, graduating in 1934. He then attended the University of California at Los Angeles (UCLA), receiving his bachelor of arts degree in mathematics in 1938, a master's degree in mathematics in 1940, and a Ph.D. in meteorology in 1946. That same year, he married Elinor Kesting Frye, a logic student and they had three children. While at UCLA, he taught physics and meteorology from 1942 to 1946. During World War II, he helped train weather officers for the armed services at the Army Air Forces Training School at UCLA.

His Ph.D. dissertation, entitled "Dynamics of long waves in a baroclinic westerly current," was published in *Journal of Meteorology* in 1947 and filled the whole issue of the journal. According to historian John Fleming, Charney's paper was influential in that it "emphasized the influence of long waves in the upper atmosphere on the behavior of the entire atmosphere rather than the more traditional emphasis on the polar front, and

it provided a simplified way of analyzing perturbations along these waves that proved both physically insightful and mathematically rigorous."
Long waves, or planetary waves, are produced
because the moving atmosphere encounters bar-
riers and is forced to ascend (based on various
obstructions such as mountains and the like) and
then descend.

Charney served for a year as a research associate at the University of Chicago for Carl-Gustaf
ROSSBY and in the academic year 1947–48 held a
postdoctoral fellowship at the National Research
Council at the University of Oslo, Norway.
According to Fleming, at Oslo he "developed a set
of equations for calculating the large-scale
motions of planetary-scale waves known as the
quasi-geostrophic approximation. Charney's technique consisted of replacing the horizontal wind
by the geostrophic wind in the term representing
the vorticity but not in the term representing the
divergence. The result was a manageable set of fil-
tered equations governing large-scale atmospheric
and oceanic flows." His landmark research was to
form the core of all theoretical work in modern
dynamical meteorology.

From 1948 to 1956, Charney served as director of theoretical meteorology at the Institute for
Advanced Study in Princeton, New Jersey. Working for John von NEUMANN, Charney was in
charge of a project to develop computer technology for weather prediction, and using ENIAC
(the Electronic Numerical Integrator and Calculator) in 1950, his team developed a successful
mathematical model of the atmosphere and
demonstrated that numerical weather prediction
was feasible. His results were the first numerical
predictions of a two-dimensional model that
approximated the actual flow at a midlevel in the
atmosphere. In 1952–53, he obtained the first
prediction of cyclogenesis (the formation of a
major low-pressure system along a baroclinic zone
or frontal boundary, with primary forcing due to
imbalances along the upper jet) with a three-
dimensional model.

In 1954, Charney helped establish a numerical weather prediction unit (Joint Numerical
Weather Prediction Unit) within the U.S.
Weather Bureau at Maryland for the routine daily
prediction of large-scale weather patterns. He
later encouraged the formation of the Geophysical Fluid Dynamics Laboratory, a research facility
utilizing computers for basic atmospheric and
oceanic research.

Charney served as professor of meteorology
and director of the Atmospheric and Ocean
Dynamics Project at the Massachusetts Institute
of Technology (MIT) from 1956 to 1981. He concentrated his research on the dynamics of the
ocean and the atmosphere, arguing that the atmosphere was a single global system that needed to
be studied. In 1966, he became the first Alfred P.
Sloan Professor of Meteorology at MIT, a chair
established through donations from A. P. Sloan,
Jr. Charney chaired the National Research Council's Panel on International Meteorological Cooperation (1963–68). In 1967, he married painter
and professor Lois Swirnoff (they divorced in
1976). From 1968 to 1971, he chaired the U.S.
Committee for the Global Atmospheric Research
Program (GARP), a decade-long international
experiment to measure the global circulation of
the atmosphere, to model its behavior, and to
improve predictions of its future state. In 1974, he
became head of MIT's meteorology department
and reorganized it into the department of meteorology and physical oceanography. In 1977, he
resigned his post as department head.

Charney was a Fellow of the American Academy of Arts and Sciences, the American Geophysical Union, and the American Meteorological
Society. He was a member of the U.S. National
Academy of Sciences and a foreign member of the
Norwegian and Royal Swedish Academies of Science. Charney's awards included the Meisinger
Award (1949), the Rossby Medal (1964), and
Cleveland Abbe Award (1980), all from the American Meteorological Society; the Losey Award of
the Institute of Aeronautical Sciences (1957); the

Symons Medal of the Royal Meteorological Society (1961); the Hodgkins Medal of the Smithsonian Institution (1968); the International Meteorological Organization prize (1971); and the Bowie Medal of the American Geophysical Union (1976).

He is known as the Father of Numerical Weather Prediction. Throughout his career, Charney's research focused on the mathematical description of large-scale atmospheric circulations that he applied to the theory of ocean currents, atmospheric wave propagation, large-scale hydrodynamic instability, hurricanes, drought, and more. He was the author of more than 60 articles and papers and mentored more than 20 M.S. and Ph.D. students.

Charney died of lung cancer in Boston on June 16, 1981. One of the supercomputers at Goddard Space Flight Center is named for him in his honor.

Cheng, Roger J.
(1929–)
Chinese
Atmospheric Scientist

Roger J. Cheng was born on June 3, 1929, in Kaifeng, along the Yellow River in the Henan Province, one of the oldest inhabited regions in China; a province of east-central China with a population of 77,130,000. During the communist revolution, it was the poorest farm country in China, and, as a result, millions of peasants died from hunger. Cheng is the son of Chen-Yu and Shin-Nan Cheng, a farming family that goes back many generations.

Fortunately for Cheng, his family's legacy of basic farm living came to an end when his father passed an exam after high school that awarded him a scholarship to college. Only two such awards are given each year by the government to Henan Province. His father went to Peking (Beijing) for higher education and became a professor at a university teaching history and political

science. He then worked for the government on land-reform policy, eventually becoming the head of the Ministry of Land Reform. Later, Henan Province elected him as a senator in the Chinese government. This assured that Cheng would not become a struggling farmer like so many of his family before him.

By passing a tough written exam, Cheng was allowed to attend the Nankai Middle School that is known for having the best and brightest students. This is the same school that many of today's top Chinese-government officers attended, including the former premier Chou En-lai. Cheng received a B.S. in physics and chemistry, but due to a disagreement with his professor over his research results, he was unable to complete his Ph.D.

As a boy, Cheng loved to take electronic devices apart and then reassemble them. He also loved photography. Both of these interests would become pivotal talents in his adult career. During one evening, a blind fortuneteller told his family that Cheng would travel beyond China, work with small things, and marry a non-Chinese woman.

After his family moved to Taiwan, Cheng passed a stringent English exam for study in America. He borrowed $50 and enough for air fare to fly to the United States, looking to further his education and opportunities in 1957. He landed in Florida and attended Florida State University for graduate work in physics and meteorology. To earn money, for three summers, he worked as a dishwasher, a waiter, and, soon after, a headwaiter at a Chinese restaurant. At the university, he became a technical assistant working for Dr. Seymour Hess, head of meteorology and an expert on the atmosphere of Mars. Cheng's early mechanical expertise came into play at Florida. He helped design very sensitive equipment to measure "water vapor" in the atmosphere of Mars in an experiment to confirm that the white caps on Mars were ice (water) and not frozen carbon dioxide. As a result of his work, he won two summer scholarships to Columbia University's Institute of Space

Physics and UCLA's Institute of Planetary Physics, the first ever for a foreign student.

While he was working at Florida State University for Hess, he read most of the papers by New York's Vincent SCHAEFER on ice-and-snow-crystal research and appreciated the work Schaefer was trying to do in cloud seeding. Many of Cheng's friends and colleagues told him that Schaefer had a reputation as a great person with whom to work and in the area of atmospheric science, at the time an emerging field.

He wrote to Schaefer, after studying everything about him, and asked for a job at the Atmospheric Science Research Center (ASRC) that Schaefer and others created in 1964 in Albany, New York. In 1966, Cheng began his first job as a technical assistant for Schaefer for $1.25 per hour, working for 10 hours a week. Cheng stayed at ASRC for 32 years.

Schaefer watched and mentored his student Cheng and, after one year, gave him an order to set up a laboratory to study and assist other ASRC scientists on the subject of atmospheric particulates. Cheng created the Laboratory for Atmospheric Particulate Analysis and was put in charge as manager.

Schaefer, who earlier had discovered cloud seeding with Irving LANGMUIR and Bernard VONNEGUT, directed Cheng to work on three projects regarding particulate matter:

1. To see particulates in terms of their sizes and shape and to study their physical property;

2. To discover their chemistry properties and what they were made of; and

3. To examine particles (aerosols) in different conditions or in three phases (vapor, liquids, and solids). In four years, Cheng developed the lab into one of the best microscopy labs, outfitted with the most modern equipment.

While working for Schaefer, Cheng made several amazing observations in three major research fields—cloud physics, marine aerosols, and environmental sciences—all later confirmed by other scientists, based on the study of single drops of water.

His first research project in the late 1960s with Schaefer, titled "Ejection of electrical charged ice particulates from a frost surface," was based on frozen water drops, ice pellets, and hail. This major discovery helped explain how the electric charges were generated in a thunderstorm cloud. Their discovery was confirmed 20 years later by scientists from MIT doing similar work. His photomicrographs that captured the activity became the featured cover in *Science Magazine* in 1970; it showed the freezing of a supercooled water drop and the ejection of microdroplets. More than 160 scientists from 25 countries requested a reprint of the photograph and article that eventually also appeared in more than 30 international science magazines, 15 encyclopedias, and a science yearbook.

Cheng's second water-drop project dealt with the formation of acid rain, marble erosion, and plant damage. Cheng demonstrated in the laboratory that flyash from electric power plants catalyzed the reaction of sulfur dioxide in water droplets to form sulfates. By studying the marble constructed city hall in Schenectady, New York, he noticed that sulfur dioxide reacted with liquid water on the surface of the stone to produce sulfuric acid. Small pollutants such as flyash act as a catalyst in the reaction, making the marble change into gypsum, which crumbles. When flyash particles are injected in water drops that are 2 to 3 millimeters in diameter and are exposed to sulfur dioxide, needlelike sulfate crystals appear. Cheng believed that this reaction is probably implicated in leaf damage. This experiment was confirmed by the Canadian EPA in 1995 who conducted similar experiments.

Finally, in 1986, he discovered with ASRC colleagues that sea-salt particles (marine aerosols) were hollow and not solid as formerly believed. The sea produces marine sea-salt aerosols by the bursting of bubbles at the surface. The small droplets are then carried high into the atmosphere by mixing and convection and can change phase to produce the sea-salt aerosol. Previous thoughts on the subject assumed they were solid and that their mass could be calculated if their diameters were known. Cheng's discovery that they are hollow makes this calculation much more difficult. Cheng's findings were confirmed by scientists from the Institute of Meteorology in 1997 in Germany. He also noticed that seawater droplets ejected sulfate particles when they changed phase; that could be important in the global sulfur cycle.

The State University of New York recognized him in 1978 with the SUNY Chancellor Award for Excellence in Professional Service. He retired from the ASRC in 2000 as Emeritus (Research scientist).

Cheng, as predicted earlier by the fortune teller, married Leida Sutt of Estonia in 1961, a non-Chinese woman; they have two sons and one daughter. Their son Mark is a computer animator whose film credits include *Dinosaur* and *Mouse Hunt*.

He has published many papers in atmospheric science and coauthored *Solar Energy Experiments for High School and College Students* (1977) with Thomas Norton and Donald Hunter and *Air Pollution and Control* in 1985. The latter book was the first of its kind in the Chinese language, was endorsed by the Director of the Chinese National Environmental Protection Bureau, and was published by their Environmental Science Service. It was distributed to all national, state, and city environmental officers as a handbook and to universities and research institutes as a reference book.

He has utilized his experience as a consultant and teacher locally in the Albany New York Medical Center, V.A. School of Microscopy, the New York State Department of Environmental Conservation, Air Resources Division, the Auto Emission Testing Lab, and others. Cheng has traveled extensively abroad especially to his homeland China, where he has given 14 seminars and visited more than 25 universities and other institutes, and to other European countries and Russia giving talks and seminars. His photomicrographs have been in demand by popular and scientific publications worldwide for their art as well as content. He is currently creating a CDROM that contains all his research papers, magazine covers, and hundreds of his photographs. Remarkably, his achievements have all centered around the study of a single drop of water.

⊠ Coriolis, Gaspard-Gustave de
(1792–1843)
French
Physicist

Born in Paris to Jean-Baptiste-Elzéar Coriolis and Marie-Sophie de Maillet Coriolis, the young man was destined to discover a force that would help shape our knowledge of weather patterns. His father fought in the American campaign with the Rochambeau Corps in 1780. Rochambeau landed in Newport with about 6,000 French troops and fought against Cornwallis at Yorktown. Jean-Baptiste became an officer with Louis XVI in 1790, but when the monarchy fell in 1792, he fled to Nancy (French province of Lorraine that borders Germany, about 2 hours from Paris). It was in that year that Gaspard was born, on May 21, 1792.

Coriolis grew up in Nancy and attended school. He went on to the École Polytechnique near Paris in 1808, graduating in highway engineering. On graduating, he entered the École des Ponts et Chaussées in the heart of Paris. He worked with the engineering corps for several years in the Meurthe-et-Moselle district near the borders of Belgium, Luxembourg, and Germany and the Vosges Mountains in the Alsace region, but when his father died, he accepted a position at the École Polytechnique in 1816.

In 1819, he introduced the terms *work* and *kinetic energy*, giving them their current scientific meaning. He proposed the *dynamode* as a unit of work, representing 1,000 kilogram-meters. He proposed the unit to measure the work of a person, a horse, or a steam engine. The term *work* stuck, but the dynamode did not catch on, as horsepower is used instead. He wanted to ensure that correct terms and definitions became part of the study of mechanics. He published his paper "Du calcul de l'effet des machines" (Calculation of Machine Effects) in 1829, later republished in 1844 as the "Traité de la mécanique des corps solides" (Treatise of the mechanics of solid bodies).

In 1829, Coriolis accepted the position of professor of mechanics at the École Centrale des Artes et Manufactures (Central School of the Arts and Manufactures) and stayed there until 1836. He also had a position at the École des Ponts et Chaussees (School of Bridges and Sidewalks) in 1832, working with Claude-Louis Navier, teaching applied mechanics.

It was in the year 1835 that he discovered the Coriolis effect or force. By studying the motion above a spinning surface on various machines, he imagined an apparent force acting on objects as they moved across the Earth's surface, due to the results of the Earth's rotation. For example, in the Northern Hemisphere, the path an object takes appears deflected to the right, while in the Southern Hemisphere it deflects to the left. This effect is responsible for wind and ocean-current patterns and is instrumental in our knowledge of weather patterns today. He published this in an 1835 paper "Sur les équations du mouvement relatif des systèmes de corps" (On the equations of the relative movement of the systems of body). During the same year, he published "Théorie mathématique des effets du jeu de billiard" (Mathematical theories related to the game of billiards).

He became a member of the Académie des Sciences in 1836. Two years later, in 1838, he became director of studies at the École Polytechnique and ended his teaching. Poor health saddled him most of his life, and he died only five years later in Paris on September 19, 1843.

Croll, James
(1821–1890)
British
Carpenter, Physicist

James Croll entered the field of science through a side door. Born in Cargill, Perthshire, Scotland, on January 2, 1821, he was the son of David Croll, a stonemason from Little Whitefield, Perthshire, and Janet Ellis of Elgin. He received an elementary school education until he was 13 years old. His knowledge of science was the result of vigilance on his part; he was self-taught. On September 11, 1848, he married Isabella MacDonald, daughter of John Macdonald.

Croll started his career far from science. A carpenter apprenticed to a wheelwright, then a joiner at Banchory, and then a shop owner in Elgin, in 1852, he opened a temperance hotel in Blairgowrie; later, in 1853, he became an insurance agent for the Safety Light Assurance Company, ending up in Leicester.

His first book, *The Philosophy of Theism*, was published in 1857 and was based on the influence of the metaphysics of Jonathan Edwards. However, due to an injury, he ended up as a janitor at Anderson's College and Museum, Glasgow, in 1859. Being a janitor gave him enough free time after his daily chores to utilize the museum's extensive library. Here, he would spend the night reading books on physics, including the works of Joseph A. Adhémar, the French mathematician, who noted in 1842 that the Earth's orbit is elliptical (having the shape of an ellipse) rather than spherical (having the shape of a sphere). Adhémar proposed in his book *Révolutions de la mer, déluges périodiques* (Revolutions of the sea, periodic floods) that the precession of the equinoxes produced variations in the amount of solar radiation striking (insolation) the planet's two hemispheres during the wintertime, and this, along with gravity effects from the Sun and the moon on the ice caps, is what produced ice ages alternately in each hemisphere, during a 26 thousand-year cycle. Precession is the slow gyration of Earth's axis around the pole of the ecliptic, caused mainly by the gravitational pull of the Sun, the moon, and the other planets on Earth's equatorial bulge. Croll also read about the new calculations of the Earth's orbit by French astronomer Urbain-Jean-Joseph Leverrier (a discoverer of the planet Neptune).

Croll decided to work on the origins of the ice age because he did not agree with the prevailing attitude that they were leftover relics from the biblical great flood, and additionally he found errors in Adhémar's work. Croll came to a different conclusion that the overriding force changing climate and creating the ice ages on Earth was variations in insolation, which is the rate of delivery of solar radiation per unit of horizontal surface, that is, the sunlight hitting the Earth.

Croll first realized that Adhémar did not take into account the shape of the Earth's orbit that varied over time and its effect on precession, and he calculated the eccentricity for several million years. He proposed that the distance between the center of an eccentric and its axis, in this case the degree of Earth's elliptic orbit, varied on a time scale of about 100,000 years. Because variations in eccentricity only produced small changes in the annual radiation budget of Earth and not enough to force an ice age, Croll developed the idea of climatic feedbacks such as changes in surface albedo (reflection) and that the last ice age ended about 80,000 years ago.

During the 1860s, he published his theories in a number of papers: "On the Physical Cause of Changes of Climate during Geological Epochs." (1864); "The Excentricity of the Earth's Orbit" (1866, 1867); "Geological Time and Date of Glacial and Miocene Periods" (1868); "The Physical Cause of the Motion of Glaciers" (1869, 1870); "The Supposed Greater Loss of Heat by the Southern Hemisphere" (1869); "Evolution by Force Impossible: A New Argument Against Materialism" (1877). During this time, he was the Keeper of Maps and Correspondence to the Scottish Geological Survey starting 1867, where he mingled with some of the best geologists of the time until he retired there in 1880.

In 1875, he published *Climate & Time in Their Geological Relations* in which he summed up his researches on the ancient condition of the Earth. On January 6, 1876, the carpenter-turned-physicist was elected a Fellow of the Royal Society of London. Among the many supporters of his nomination, as was customary to be elected, was Charles Darwin. Croll received an L.L.D. (law degree) that year from St. Andrews College. Although his main interests were in the field of paleoclimate change, he also put forth theories about ocean currents and their effects of climate during modern times.

However, some of his thoughts and ideas were wrong; for example, Croll believed that ice ages varied in the hemispheres, and his estimated age for the last ice advance ending 80,000–100,000 years ago was wrong, when in actuality it ended between 14,000 and 10,000 years ago, as research currently shows. Because of these errors, Croll fell out of vogue until 1912 when Yugoslav geologist Milutin Milankovitch revised Croll's theories in his book, *Canon of Insolation*.

Croll published close to 90 papers on a variety of subjects, such as "Ocean Currents" (1870; 1871; 1874); "Change of Obliquity of Ecliptic: its effect on Climate" (1867); "Physical Cause of Submergence during Glacial Epoch" (1866; 1874); "Boulder Clay of Caithness & Glaciation of North Sea" (1870); "Method of determining mean thickness of Sedimentary Rocks" (1871); and "What Determines Molecular Motion?— The Fundamental Problem of Nature" (1872).

A famous debate between Croll and Irish scientist William Carpenter on the deep-sea circulation during the 1860s–1880s was well discussed in the literature and around scientific circles via correspondence. In 1885, Croll published *Climate and Cosmology* to answer critics of his earlier work *Climate & Time in Their Geological Relation*. Five years later, plagued by ill health his whole life, he died in Perth on December 15 at age 69, shortly after publishing a small book called *The Philosophical Basis of Evolution*.

D

Dalton, John
(1766–1844)
British
Chemist, Physicist

John Dalton was born September 5, 1766, in Eaglesfield, Cumberland (now Cumbria), England, the youngest of three children, to Joseph Dalton and the former Deborah Greenup, both Quakers. John and his brother went to the Quaker school at Pardshow Hall, and it was here that he became proficient in mathematics. It was the mentoring of a wealthy Quaker, Elihu Robinson, who stimulated his interest in science, especially meteorology. Robinson was a friend of the blind philosopher John Gough, who also befriended Dalton and encouraged him to keep a daily journal. Dalton began to compile a daily meteorological diary in 1787 that he maintained for 46 years until his death; it is reported to have contained 200,000 entries, much of them relating to the subject of gases.

At only 12 years of age, he became a schoolmaster until the school closed in 1780; then he worked temporarily on a farm owned by an uncle. He joined his brother, also a teacher, in 1781 who was working at their cousin George Bewley's school in Kendal. When Bewley retired, John and his brother reopened the school from 1785 to 1793. In 1793, Dalton moved to Manchester and was professor of mathematics and natural philosophy at New College. Here, he joined the Manchester Literary and Philosophical Society and published his first book in 1793, *Meteorological Observations and Essays,* in which he correctly concluded that the aurora borealis is a magnetic phenomenon. Other meteorological contributions included observations on the trade winds, and the cause of rain; he determined the point of the maximum density of water, explained the condensation of dew, and developed a table of vapor pressures of water at various temperatures. In 1794, he was the first to provide a scientific description of color blindness, a condition from which he and his brother suffered and which was long after called Daltonism. DNA extracted from one of his preserved eyes showed that he lacked the pigment that gives sensitivity to green, the classic condition known as a deuteranope.

When New College moved to York in 1799, Dalton stayed in Manchester to offer private lessons in mathematics and to perform his own chemistry experiments. A keen observer, he kept tables of barometric pressure, temperature, wind, humidity, and rainfall by using self-made meteorological instruments. His observations led him to think about the composition of air and the behavior of gases, and he concluded that the atmosphere was a combination of independent gases and not a "solution" of them.

In 1802, he presented his paper, "Experimental Essays on the Constitution of Mixed Gases; on the Force of Steam or Vapour from Water and Other Liquids in Different Temperatures, Both in a Torricellian Vacuum and in Air; On Evaporation; and on the Expansion of Gasses by Heat," to the Manchester Literary and Philosophical Society. This landmark work established "Dalton's law of partial pressures"; that is, to say the total pressure of gas is equal to the sum of the partial pressures of the individual component gases. This was based on his study of vapor pressures over a liquid and their accumulative effect of the pressures of individual constituents in a combination of various gases. Dalton's equation for the evaporation rate, based on vapor pressure and wind velocity as described in the 1802 paper, is still employed by meteorologists. Dalton also discovered that gases expand as their temperature rises, and he demonstrated the solubility of gases in water, the rate of diffusion of gases, and the constancy of the composition of the atmosphere.

On October 21, 1803, Dalton read his paper "On the Absorption of Gases by Water and Other Liquids" before the Literary and Philosophical Society of Manchester. This paper also included his first periodic table of 21 relative atomic and compound weights; it was the beginning of his atomic theory of matter that all elements are composed of tiny, identical, and indestructible particles that he later described in the first volume of his *New System of Chemical Philosophy* (2 vols., 1808–27). This volume is considered by some to be one of the most important scientific books of the 19th century. Here he also proposed (1) standard symbols for the elements, (2) the beginning of a system of standard chemical notation, and (3) development of the laws of definite and multiple proportions. Dalton also used ball-and-stick models to represent atoms and molecules in three-dimensions and is considered the world's first stereochemist, although in print he referred to molecules in two-dimensional terms.

He was president of the Manchester Philosophical Society from 1817 to his death, and he contributed most of his writings to it. He also became a member of the Royal Society in 1822 and in 1825 received a medal for his work on the atomic theory. In 1830, the French Académie des Sciences (Academy of Sciences) elected him as one of their eight foreign associates. The following year, he was active in founding the British Association for the Advancement of Science and was given a pension by the association, starting in 1833. The unit for atomic weight was called a dalton for many years, a name still used today in biochemistry. Dalton is best known as one of the fathers of modern physical science and the father of the Atomic Theory.

In 1837, Dalton had a stroke that partially disabled him. He suffered a second stroke in May 1844 and died on July 24, shortly after noting the weather conditions for the day in his journal.

Manchester buried him like a king—the opposite of his Quaker lifestyle—with his body lying in state and a royal funeral; the funeral procession was more than a mile long. More than 400,000 people viewed his body as it lay in state. Furthermore, one of Manchester's great Town Hall murals contains one image of John Dalton collecting marsh gas, painted by the famous pre-Raphaelite painter Ford Madox Brown, as well as a statue of him on the porch of town hall by Victorian sculptor Sir Francis Chantrey.

Daniell, John Frederic
(1790–1845)
British
Chemist, Meteorologist

John Frederic Daniell was born in London, England, on March 12, 1790, the son of a lawyer. Apparently, he was privately educated but later attended Oxford University, graduating with a strong technology background, although historians are not sure if a degree was honorary or earned. He continued his education and attended lectures in 1812 at the medical school on Windmill Street, London, delivered by William Thomas Brande, professor of chemistry at the Royal Institution. Daniell was elected a Fellow of the Royal Society, London, in 1814, mostly as a result of Brande and others who promoted him and who were impressed with his early work in both the refinery business and his geologic interests.

Daniell's interest in geology included maintaining a large collection of rocks and minerals that he gathered. In 1815, he embarked on a geological tour of the British Isles with Brande and the following year coauthored *A Descriptive Datalogue of the British Specimens Deposited in the Geological Collection of the Royal Institution* (London, 1816). Besides the interest in geology, Daniell was experimenting in meteorology while living at his father's home in Lincoln Inns Fields. By 1820, he invented a dew-point hygrometer, a device that measures relative humidity and is now a standard instrument in meteorological observation. This Daniell hygrometer was made up of two thin glass bulbs hanging from a base and joined with a glass tube. He first described it in an article, "On the New Hygrometer," that was published in the *Quarterly Journal of Sciences, Literature, and the Arts*. With the hygrometer, it was easy to determine both the absolute and the relative humidity, then known as the expansion corresponding to the dew point and the relative expansion.

He then published *Meteorological Essays and Observations* in 1823, an important book that discussed meteorological effects of solar radiation and the cooling of the Earth and was one of the first explanations of the trade winds. To help improve the status of British meteorology, a group led by physician George Birkbeck met at the Ludgate (London) Coffee House on October 15, 1823, and founded the Meteorological Society of London. Daniell, Luke HOWARD, and Thomas Ignatius Maria Forster were among its council members. However, for unknown reasons, it met for only several months, yet stayed in existence until 1836 when it was revived, but it failed again in 1843. In 1848, another Meteorological Society of London was founded; yet it too died a quick death. By 1850, the British Meteorological Society was formed and took possession of much of the material from the previous attempts to create a meteorological society.

In 1824, Daniell's knowledge of the atmosphere was applied to horticulture and led to improvements in the growth of hothouse plants. His "Essay on Artificial Climate Considered in Its Applications to Horticulture" showed the importance of humidity in greenhouses and earned him the silver medal award from the Horticultural Society in 1824. The following year, Daniell held a four-year job as director of the Imperial Continental Gas Company and toured France and Germany to promote gas lighting. As a result, gas lighting first started in Germany in the city of Hannover that year.

In 1826, Daniell was one of the founding members of Society for the Diffusion of Useful Knowledge, along with attorney Baron Henry Brougham and publisher/journalist Charles Knight. This short-lived organization was dedicated to publishing journals, maps, and other information for all classes of English readers. He went on to invent a modified pyrometer, a type of thermometer that was successful in measuring internal heat, improving on the one invented by Wedgewood (used for checking his kiln temperature). Daniell also invented a version of a water barometer in 1830.

Daniell was hired as the first professor of chemistry and meteorology at the new King's College in 1831 (a post that lasted to 1845) and established a department of applied science. King's College was located next to Somerset House on land granted in perpetuity by the Crown in the heart of London in 1829. The college was founded by royal charter with the support of King George IV, the Duke of Wellington, and a distinguished group of patrons, both ecclesiastical and secular, who wished to promote a university education that encompassed both traditional and modern studies but was less radically secular than the institution that is now University College.

In 1832, Daniell was awarded the Rumford Medal of the Royal Society of London for his earlier work with the pyrometer; yet, he never gave up his interest in rocks and minerals. During 1833, he collaborated with England's most famous mineralogist William Hallowes Miller (1801–80), professor of mineralogy at the University of Cambridge, on the subject of spectra, the distribution of energy emitted by a radiant source. Miller demonstrated that when sunlight is passed through gases in the laboratory, additional dark lines appear in the sun's spectrum and suggested that the dark lines were due to the presence of gases on the sun. Miller became the first to take photographs of spectra.

Daniell began a long collaboration in 1824 with his friend Michael Faraday, who was also mentored by William Thomas Brande. Thanks to Faraday, Daniell became interested in electrochemistry and turned his attention to the creation of batteries.

Daniell is best known for creating the Daniell constant cell, an early battery that he developed between 1835 and 1837 and that became the first reliable source of DC current. It was very popular with telegraph operators in Britain, America, and France due to its reliability. He was awarded the Copley Medal by the Royal Society in 1836. That same year, he served as a member of the committee of the Royal Society on behalf of the admiralty to standardize meteorological observations throughout the British Empire, proving that he never left his interest in meteorology. He also taught chemistry at the Military School of the East India Company, in Addiscombe, Surrey, during this time (1835–44).

Daniell published his monograph *An introduction to the study of chemical philosophy, being a preparatory view of the forces which concur in the production of chemical phenomena* in 1839 and dedicated it to his friend Faraday. During his career, Daniell served as the foreign secretary for the Royal Society (1839–45) and was a member of the Admiralty Commission on Protecting Ships from Lightning (1839). In 1841, he became a founding member and vice president of the Chemical Society of London. He wrote many papers that were published in various journals, including *On a new hygrometer* (1820), *Chemistry* (1829–38), *On voltaic combinations* (1836), and *On the spontaneous evolution of sulpheretted hydrogen in the waters of the western coast of Africa* (1841).

Daniell was married and the father of five daughters and two sons. He had a heart attack at the Royal Society on March 15 1845, and died at the age of 55.

⊠ **Day, John**
(1913–   )
American
Meteorologist

John Day was born in Salina, Kansas, on May 24, 1913, the older of two sons of Arthur Cutler Day and Lenora (Wilson) Day. After briefly residing in Salina and in Pocatello, Idaho, the family moved to Colorado Springs, Colorado, where he lived until 1936. He experienced a normal well-nurtured boyhood with strong interest in sports (baseball, tennis, golf, and basketball). Many happy hours were spent at the family cabin at nearby Carroll Lakes, a flyfishing club, located at 9,000 feet on the south slopes of Pikes Peak.

His strong-willed mother started him on piano lessons at age 6. This was to stand him in good stead in later years, as this skill was to be the means of paying his college tuition. High school activities, beyond the classroom, were focussed on basketball and tennis and band (clarinet). Summer money was earned at the local golf course, where he caddied and learned to play the game.

In 1931, in the middle of the Great Depression, Day began his studies in physics at the local and inexpensive Colorado College. He worked his way through school as a pianist for Johnny Metzler's orchestra, playing engagements at local hotels. Because of his preoccupation with music, he was only an average student.

He graduated in 1936 from Colorado College with a B.A. in physics and enrolled that summer at the Boeing School of Aeronautics in Oakland, California. This period saw the beginnings of the airline industry. There was great need for trained personnel, and, though expensive, Boeing School was guaranteeing employment in the industry.

Arriving at the school, he found that a new nine-month curriculum in airline meteorology had just been opened and that the school was looking for students with backgrounds in physics and mathematics. He was a ready and willing recruit. In 1937, he landed a job at Pan American Airways, which had just started to operate its "flying boats," the four-engine Clippers, from the former naval base. He began part time as an airport guide, dealing with Clipper arrivals and departures, which drew thousands of spectators, then worked as a mail clerk, and finally became an assistant in the meteorology office. Here, he plotted meteorological data on maps that were to be analyzed later by senior meteorologists who provided flight crews with wind and weather-route forecasts.

Pam Am's venture in flying across thousands of miles of the North and South Pacific Ocean was a monumental undertaking. It required establishment of landing sites, sometimes only specks of coral atolls, as in Canton Island, located just south of the equator. This required setting up a communications network because there was none. Likewise, it required the establishment of meteorological observing stations, for there were none of these either. It required construction of a fleet of four-engine flying boats from the Martin and Boeing aircraft companies, serviced by ground personnel, including meteorologists.

After graduating from Boeing School in 1937 with an aeronautics certificate of competence in airline meteorology, Day accepted employment in the meteorology department of Pan American Airlines. After a two-year stint in Alameda, he was sent to the office in Pearl City, Hawaii, and then in 1941 he was transferred to head the Auckland, New Zealand, office.

During World War II, Pan Am was taken into the navy as a transport wing, and Day was

John Day. Day, along with coauthor Vincent Schaefer, is the author of the popular *Peterson's Field Guide to the Atmosphere*. His most recent book is *The Book of Clouds*. (Courtesy of John Day)

commissioned a lieutenant, junior grade. He was part of a pioneering party to extend flight service to Australia and was assigned to duty at Brisbane. On termination of this assignment, he returned to duty at the Fleet Weather Center in Honolulu. No longer under the auspices of the navy, Pan Am assigned Day in 1946 to go from Manila to Tokyo to provide forecasting services for a special project, the transport of UNRRA (United Nations Relief and Rehabilitation Administration) personnel from the United States to China. The chosen route was the great circle along the Aleutians to Tokyo, previously unflown, though today this route is the expressway of flights to the Orient. In 1946, this was a pioneering venture through a very weather-active region. Forecasts had to be made from a very sparse network of observing stations in mainland Asia and over the western Pacific Ocean.

Returning to the United States in 1947, Day submitted his resignation to Pan Am to pursue a career in academia and found a position at Oregon State University's department of physics. Two other ex-meteorologists were hired, so the three of them informally started a subdepartment of meteorology, which later grew to become an independent department of atmospheric science. While teaching engineering physics and meteorology at Oregon State University, he undertook an assault on the doctorate degree there, earning an M.S. in 1952 and completing the Ph.D. in 1956 in cloud physics with a dissertation "Nucleation of Supercooled Water Drops."

That same year, 1956, he obtained a two-year appointment at the University of Redlands (60 miles east of Los Angeles), and returned to Oregon in 1958 to take a position at Linfield College.

In 1962, Day was awarded a science faculty fellowship by the National Science Foundation to study and do research in Dr. B. J. Mason's cloud-physics group at Imperial College of Science and Technology in London. He continued his interest in the generation of cloud-condensation nuclei by the bursting of air bubbles at the surface of the ocean. Cloud-condensation nuclei (CCN) are small particles in the atmosphere about which water vapor may condense to form clouds. Through attendance at cloud-physics conferences around the world, Day broadened his friendship with other researchers in the field, among them Vincent J. SCHAEFER and Duncan BLANCHARD.

In 1971, Day was awarded a sabbatical from Linfield and returned to England, where he was headquartered at the Meteorological (Met) Office at Bracknell; from this base, he studied the history of clouds and cloud atlases, becoming intimately acquainted with Luke Howard, the "Godfather of Clouds."

That year, he received a letter that was to shape his life from Vincent Schaefer, inviting him to collaborate in writing a *Peterson's Field Guide to the Atmosphere* (1981), a long-range project. It was followed by a second book in the well-known series *A First Guide to Clouds and Weather* (1991). Other writing projects in the 1970s and 1980s produced three meteorology textbooks and a book on his favorite topic, *Water, the Mirror of Science* (1961, with Kenneth Davis) and, most recently, in 2002, a book on clouds titled *The Book of Clouds*. Previous books include *Rudiments of Weather* with Fred W. Decker (1955), *The Science of Weather* (1966), and *Climate and Weather* with Gilbert Sternes (1970).

From 1973 to 1978, he initiated and directed Linfield's new venture in adult off-campus programming. For this work, he was honored in 2000 by being selected for membership in the Tall Oaks Society for having rendered outstanding service to the college. He retired from Linfield in 1978.

In the 1980s, cloud photography became a serious hobby and even a vocation for Day. Operating under the name of Dayphoto Pacific, he built up his collection of cloud images, sold a line of photographic artcards, and had exhibitions of cloud images in various galleries, including the Brooks Institute of Santa Barbara and a permanent exhibit at the Science Museum in Hong Kong. These were to become the backbone of a website (www.cloudman.com) developed in 1997 by one of his daughters.

Because of his love of the subject, Day has continued to teach meteorology since retirement. He has provided community service in the form of education in meteorology through a weekly column, "Words on the Weather," written in the *News Register* of McMinnville, Oregon, since 1977. He considers his contributions to the subject of meteorology to be primarily as an educator, secondarily as an author, and finally as a photographer of clouds. He has been married to Mary Hyatt for 65 years and has five children. He continues to teach and comment on the field of weather and climate through his website.

Emanuel, Kerry Andrew
(1955–   )
American
Meteorologist

Kerry Emanuel was born on April 21, 1955, in Cincinnati, Ohio, to Albert Emanuel II, a mechanical engineer and consultant, and Marny Schonegevel Emanuel, a former flight instructor. His early education was scattered between Montgomery Country Day School, Wynnewood, Pennsylvania; Palm Beach Day School, Palm Beach, Florida; and Deerfield Academy, Deerfield, Massachusetts; his interests included classical music, weather, and sailing. He attended Boston's Massachusetts Institute of Technology (MIT), where he received a Bachelor of Science degree in earth and planetary sciences in 1976 and a Ph.D. in meteorology two years later.

In 1989, Emanuel became professor and director of the Center for Meteorology and Physical Oceanography in the Department of Earth, Atmospheric, and Planetary Sciences at MIT. The following year, he married Susan Boyd-Bowman. They have one son. Beginning in 1997, he was professor in the department's program in atmospheres, oceans, and climate.

Emanuel has discovered a general criterion for slantwise moist convection in the atmosphere and developed a quantitative upper-bound theory for tropical cyclone intensity. This has increased understanding and ability to forecast mesoscale precipitation bands in cyclones and to forecast hurricane intensity change. This is a major contribution in meteorology as it gives a quantitative understanding of hurricane intensity and structure. He is nationally recognized for his work and has hypothesized the possibility of "hypercanes," extreme forms of hurricanes that may have supersonic winds and extend upward into the middle stratosphere, the region of the atmosphere above the troposphere and below the mesosphere. These storms may have occurred following asteroid impacts with the ocean or after extensive undersea volcanism.

He has produced close to 100 publications and has been the recipient of the American Meteorological Society's Meisinger Award (1986) and Banner I. Miller Award (with Richard Rotunno) in 1992. He was elected a Fellow of the American Meteorological Society in 1995.

Emanuel's current research interests include air-sea interaction at very high wind speeds and the possibility that global tropical cyclone activity drives the oceans' thermohaline circulation and thereby has a strong feedback on climate. The density of seawater is controlled by its temperature (*thermo*) and its salinity (*haline*), and the circulation driven by density differences is thus called the thermohaline circulation.

⊠ Espy, James Pollard
(1785–1860)
American
Meteorologist, Mathematician

James Pollard Espy was born in Westmoreland County, Pennsylvania, on May 9, 1785. He obtained a degree at Transylvania University in Lexington, Kentucky in 1808 and became principal of the classical academy in Cumberland. He later studied law and was admitted to the bar, practicing in Xenia, Ohio. In 1817, he accepted a position in the classical department at the Franklin Institute in Philadelphia and began to write for their journal on meteorological subjects.

He was perhaps the first to describe how clouds form. In addition, Espy promoted a convection theory of storms in 1836 before the American Philosophical Society, where he was awarded their Magellan gold medal. Chairman of a joint committee of the American Philosophical Society and the Franklin Institute, he investigated data on storms; their annual reports were published from 1834 to 1838. In 1840, he also promoted his theories before British and French scientific societies and contributed papers to their proceedings.

To combat droughts, he unsuccessfully tried to make it rain, believing that fires that generated strong upward currents could produce rain. In the 1830s and 1840s, Espy and naturalist William C. Redfield publicly clashed on the origin of storms (the American Storm Controversy) when Redfield's findings were first published in *The American Journal of Science* in 1831. Both were right and wrong, and the debate, which peaked from 1834 to 1843, stimulated many in the field of meteorology. Espy published *Philosophy of Storms* in 1841 and, as a result, was dubbed the "Storm King."

The following year, he became a meteorologist to the War Department (and the navy department in 1848). While there, Espy developed the use of the telegraph in assembling daily weather observation data and studied the progress of storms; he also detected the traits of cyclones and tornadoes and laid out the framework for scientific weather forecasting. This work was championed by Joseph HENRY in the Smithsonian in 1846 until the Civil War and was later implemented by Cleveland ABBE (1870) at what would become the U.S. Weather Bureau. Espy died in Cincinnati, Ohio, on January 24, 1860.

F

Fahrenheit, Daniel Gabriel
(1686–1736)
German
Instrument Maker, Physicist

Daniel Gabriel Fahrenheit, a German instrument maker and physicist, was born in Danzig, Germany (now Gdansk, Poland) in 1686, the oldest of five children. Fahrenheit's major contributions lay in the creation of the first accurate thermometers in 1709 and a temperature scale in 1724 that bears his name even today.

When he was 15 years of age, his parents died of mushroom poisoning. As orphans, the four younger Fahrenheit children were placed in foster homes, and Daniel was apprenticed to a merchant who taught him bookkeeping. He was sent to Amsterdam circa 1714, where he learned of the Florentine thermometer, invented in Italy 60 years prior in 1654 by Grand Duke Ferdinand II of Tuscany (1610–70), one of the great family of the Medici. For some unknown reason, his curiosity was sparked, and he decided to make thermometers for a living. He abandoned his bookkeeping apprenticeship, whereby Dutch authorities issued warrants for his arrest. While on the run, Fahrenheit spent several years traveling around Europe and meeting scientists, such as Danish astronomer Ole Romer. Eventually, he came back to Amsterdam in 1717 and remained in the Netherlands for the rest of his life.

What seems so simple today—having a fixed scale and fixed points on a thermometer—was not worked out for a long time because several makers used different types of scales and liquids for measuring. In 1694, Carlo Renaldini, a member of the Academia del Cimento and professor of philosophy at the University of Pisa, was the first to suggest taking the boiling and freezing points of water as the fixed points. The academy was founded by Prince Leopoldo de' Medici and Grand Duke of Tuscany Ferdinand II in 1657 with the purpose to examine the natural philosophy of Aristotle. The academy was active sporadically for 10 years and concluded its work in 1667 with the publication of the *Saggi di Naturali Esperienze (Essays of Natural Experiments made in the Academie del Cimento)*.

Unfortunately, Florentine thermometers—or *any* thermometer of the time—were not very accurate; no two thermometers gave the same temperature because there was no universal acceptance of liquid type or agreement on what to use for a scale. Makers of Florentine thermometers marked the lowest scale as the coldest day in Florence that year and the hottest day for the highest scale. Because temperature fluctuations naturally occur over the years, no two thermometers gave the same temperature. For several years, Fahrenheit experimented with this problem, finally devising an accurate alcohol thermometer in 1709 and the first mercury, or "quicksilver," thermometer in 1714.

Fahrenheit's first thermometers, from about 1709 to 1715, contained a column of alcohol that directly expanded and contracted, based on a design made by Danish astronomer Ole Romer in 1708, which Fahrenheit personally reviewed. Romer used alcohol (actually wine) as the liquid, but his thermometer had two fixed reference points. He selected 60 degrees for the temperature of boiling water and 7.5 degrees for melting ice.

Fahrenheit eventually devised a temperature scale for his alcohol thermometers with three points, calibrated at 32 degrees for freezing water, 96 degrees for body temperature (based on the thermometer being in a healthy man's mouth or under the armpit), and zero degrees fixed at the freezing point of ice and salt, believed at the time to be the coldest possible temperature. The scale was etched in 12 major points, with zero, four, and twelve as the three points and eight graduations between the major points, giving him a total of 96 points for his scale for body temperature on his thermometer.

Because his thermometers showed such consistency among them, mathematician Christian Wolf at Halle, Prussia, devoted a whole paper in an edition of *Acta Eruditorum*, one of the most important international journals of the time, to two of Fahrenheit's thermometers that were given to him in 1714. From 1682 until it ceased in 1731, the Latin *Acta Eruditorum*, published monthly in Leipzig, and supported by the Duke of Saxony, was one of the most important international journals. The periodical was founded by Otto Mencke, professor of morals and practical philosophy, and mathematician Gottfried Wilhelm Leibnitz. Written in Latin, the journal covered science and social science and was primarily a vehicle for reviewing books. In 1724, Fahrenheit published a paper "Experimenta circa gradum caloris liquorum

nonnullorum ebullientium instituta" (Experiments done on the degree of heat of a few boiling liquids) in the Royal Society's publication *Philosophical Transactions* and was admitted to the Royal Society the same year.

Fahrenheit decided to substitute mercury for the alcohol because its rate of expansion was more constant than that of alcohol and could be used over a wider range of temperatures. Fahrenheit, like Isaac NEWTON before him, realized that it was more accurate to base the thermometer on a substance that changed consistently based on temperature, not simply on the hottest or coldest day of the year like the Florentine models. Mercury also had a much wider temperature range than alcohol. This was contrary to the common thought at the time, promoted by Halley as late as 1693, that mercury could not be used for thermometers due to its low coefficient of expansion.

Fahrenheit later adjusted his temperature scale to ignore body temperature as a fixed point, bringing the scale to just the freezing and the boiling of water. When he died, scientists recalibrated his thermometer so that the boiling point of water was the highest point, changing it to 212 degrees as Fahrenheit had earlier indicated in a publication on the boiling points of various liquids, and the freezing point became 32 degrees (body temperature became 98.6 degrees). This is the scale that is presently used in today's thermometers in the United States and some English-speaking countries, although most scientists use the Celsius scale.

By 1779, there were some 19 different scales described in use on thermometers, but it was Fahrenheit, astronomer Anders CELSIUS, and Jean Christin, whose scales were presented in 1742 and 1743, who finally helped to set the standard for an accurate thermometer, standards that are being used today. Besides making thermometers, Fahrenheit was the first to show that the boiling point of liquids varies at different atmospheric pressures and who suggested this as a principle for the construction of barometers. Among his other contributions were a pumping device for draining the Dutch polders and a hygrometer for measuring atmospheric humidity.

On September 16, 1736, at the age of 50, Fahrenheit died at The Hague. There is virtually no one in the English-speaking countries today who does not have a thermometer with his initials on it.

Fitzroy, Robert
(1805–1865)
British
Meteorologist

Robert Fitzroy, son of General Lord Charles Fitzroy and fourth great grandson of Charles II, was born on July 5, 1805, at Ampton Hall, Suffolk, England. He attended the Harrow boarding school and then the Royal Naval College at Portsmouth, where he excelled as a student. He entered the Royal Navy in October 19, 1819, and was commissioned September 7, 1824. It wasn't until he retired from the navy in 1850 that he turned his attention to meteorology, and he made significant contributions before his death.

Shortly after his naval commission, Fitzroy embarked on a surveying expedition to explore the coasts of Patagonia and Tierra del Fuego, South America (1828–30). He took command of the ship, the HMS *Beagle* after the captain committed suicide, ironically to be the fate of Fitzroy years later. Fitzroy completed the surveying mission of the coasts of Patagonia, Tierra del Fuego, and the Strait of Magellan. A second surveying mission went to the same region from 1831 to 1836 and was arranged by naval hydrographer Admiral Francis BEAUFORT. The *Beagle*, loaded with many meteorological instruments including 24 chronometers and barometers, was the first voyage with orders that wind observations were to take place using the Beaufort wind scale. The Royal Navy's Beaufort developed a scale in 1805, the wind-force scale, to estimate wind speeds from effects on ships' sails.

The second cruise was to chart the coasts of South America and to secure longitudes by chronological measurements. However, Fitzroy hoped that the voyage would substantiate the Bible's Book of Genesis and the story of the flood (finding evidence that it indeed covered the world). A young naturalist, Charles Darwin, was recommended to go on the voyage with Fitzroy by Cambridge's professor of botany, John Stevens Henslow. Darwin was the naturalist passenger as they circumnavigated the Earth. In 1839, they cowrote the three-volume *Narrative of the Surveying Voyages of His Majesty's Ships Adventure and Beagle* that discussed the findings of the *Beagle*'s two voyages. Darwin wrote volume three, which he later updated and published as the popular *Voyage of the Beagle*. It was during this voyage that Darwin formulated theories on evolution that disagreed with Fitzroy, the creationist and religious fundamentalist. It is thought to be part of the reason why Fitzroy killed himself.

In 1837, Fitzroy was awarded the gold medal of the Royal Geographical Society for his surveying expedition and entered politics shortly after his voyages. In 1841, he served as a Member of Parliament for Durham and was governor general of New Zealand from 1843 to 1845. He became very unpopular while defending the rights of New Zealand Maoris, the native population of Polynesian-Melanesian descent, to claim land and so was recalled. He resigned from the navy in 1850, although he was given the rank of rear admiral in 1857 and vice-admiral in 1863.

Elected a Fellow of the Royal Society in May 1851 with Beaufort and Darwin as two of the sponsors, in 1854 Fitzroy began work for the meteorological department of the British Board of Trade and became the first director of the new Meteorological Office that was formed in 1855. This bureau, known as the Met Office, may be the first weather office in the world still in existence and was created to compile statistics and charts on wind to aid in maritime navigation.

Beginning in 1860, Fitzroy prepared the first daily weather forecasts in England that were published in *The London Times*, although these drew some opposition from scientists who believed that you could not predict the weather, agreeing with other newspapers and even Parliament. Capitalizing on the invention of the telegraph, he set up a telegraph network for the rapid collection and dissemination of meteorological observations, also in 1860. His telegraph forecasts were posted in Lloyd's of London, where sailors and captains mingled before sailing. America's Joseph HENRY started to set up such a network in 1847. Eleven British stations telegraphed meteorological conditions to Fitzroy, and he became one of the first to attempt to forecast the weather scientifically. He is credited with first using the term *forecast*.

Several types of barometers bear his name, although there is no evidence that he invented them all. One that Fitzroy did design was specifically for sailors to use before setting sail and is credited for saving many lives: Fitzroy barometers often included thermometers and other gadgets as well as instructions for interpretation. His barometers are still available and used today.

Fitzroy also was a pioneer in producing weather charts and introduced symbols for wind speed and direction, pressure, and temperature. The term *synoptic chart* is attributed to him, and his charts were used to produce gale warnings for shipping. *The Weather Book*, his only major book, was published in 1863 and contains pictures of storms and clouds and diagrams and charts.

Historians suggest that both his assistance of Darwin, which greatly troubled him later because of his evolution theory, and criticism by his colleagues over forecasting abilities led him to take his own life on April 30, 1865, at his home outside London. He is buried in the Churchyard of All Saints church in Upper Norwood, London, England. After his death, the Meteorological Department was closed; however, by popular demand, forecasting resumed in 1867, ironically

with the help of Charles Darwin's half cousin, Francis GALTON.

Fitzroy Island in the Galápagos is named for him. In Australia, John Clement Wickham, a member of the HMS *Beagle*, named the Fitzroy River for him on February 26, 1838. Fitzroy Crossing is also named for him.

⊠   **Franklin, Benjamin**
(1706–1790)
American
General Science, Physics, Oceanography

Benjamin Franklin is famous as one of early America's premier diplomats, but in between his times helping to nurture democracy in a young republic, he made important contributions to meteorology. Born in Boston, Massachusetts, to tallowmaker Josiah Franklin and his second wife Abiah Folger, Franklin was the youngest addition to a family of 11 older brothers and sisters. He attended South Grammar School at the age of eight and then the Brownell School for Writing and Arithmetic (1715–16). This was the end of his formal schooling.

Before Franklin entered the world of science and politics, he worked in the printing business for more than a decade. He was first apprenticed to his father and then sent to his domineering half-brother James, a printer, in 1718. In 1721, James started a newspaper called the *New England Courant*, a "lively and irreverent" newspaper that was the first to feature humorous essays and other literary content. James made Franklin its editor, and in his brother's paper, he wrote the country's first essay series (in 14 installments) called "Silence Dogood." He submitted the series anonymously, fearing that his brother would not publish the essays otherwise.

In 1723, Franklin moved to Philadelphia. He was penniless but found a job with printer Samuel Keimer. He sailed for England in 1724 to purchase materials for a printing shop at the advice of Pennsylvania Governor Keith, who volunteered to loan Franklin the money to set up his own press. Instead, Franklin worked in London for two years as a printer and returned to Philadelphia and, by 1727, returned to work for his old boss Keimer. Shortly after, he started his own printing business with partner Hugh Meredith. He founded the "Junto" or "Leathern Apron" Club, a Friday night affair for self-improvement that lasted until 1765. In 1728, Franklin and Meredith bought the failing *Pennsylvania Gazette* from old boss Keimer and kept publishing.

Franklin also found the time to marry Deborah Read in 1730 and had a son (who died at age four of smallpox) and a daughter Sarah. He already had an illegitimate son William, although some historians attribute William to Deborah. Franklin and Meredith were appointed Pennsylvania's official government printers, but Franklin bought him out in July. The following year, he founded the Library Company (now Philadelphia Library) and was its director and librarian until 1733.

The publication that would give Franklin fame and wealth for some 25 years began in 1732 when he published the first volume of his famous *Poor Richard's Almanack*. His first recorded weather forecasts were published in this almanac. The almanac was full of weather information, advice for farmers, tips for good health, and proverbs for living that are still quoted today.

In 1737, Franklin bought a "pocket compass & microscope" (which may indicate when his interests turned to science) and became Philadelphia's postmaster. His inventive mind turned to heating, and in 1741, he invented the open or "Franklin" stove, also referred to as the "Pennsylvania fireplace." He never obtained a patent on this or other inventions, believing that his inventions should be in the public interest.

On May 14, 1743, Franklin published *A Proposal for Promoting Useful Knowledge Among the British Plantations in America*, in which he wrote, "The first drudgery of settling new colonies is now pretty well over, and there are many in every

Benjamin Franklin. World famous for his lightning experiments using kites, Franklin also invented the lightning rod. *(Original mural by Thorton Oakley, Photo by J. J. Barton, The Franklin Institute, Courtesy AIP Emilio Segrè Visual Archives)*

province in circumstances that set them at ease, and afford leisure to cultivate the finer arts, and improve the common stock of knowledge." With that said, he founded the American Philosophical Society in 1744. By 1769, it had national prestige and included as members George Washington, John Adams, Thomas Jefferson, Alexander Hamilton, Thomas Paine, Franklin Rush, James Madison, and John Marshall.

It is not known if Franklin had an interest in meteorology during this time, but the year 1744 is the first year that *meteorologist* can truly be attached to him. He attended Archibald Spencer's lectures in natural philosophy, which fueled his interest in electricity. A large early evening storm on Friday, October 21, prevented Franklin from watching a lunar eclipse. However, when he received copies of the northern newspapers, he read that the eclipse had been observed in Boston and other places to the northeast. As early as 1734, Franklin was tracing the courses of thunderstorms and published it as a "Note on a Thunderstorm," but he began to gather information on the progress of hurricanes and found that they all moved up from the southeast. He suggested that if weather observers could follow the storms, they could notify people ahead of the path. More than 100 years later, Joseph HENRY did just that.

In April 1745, Franklin's electricity experiments began in earnest after friend Peter Collinson sent a glass tube and a pamphlet describing recent German electrical experiments to the Library Company. He turned his house into an electricity laboratory and quickly learned firsthand the effects of electricity while getting shocked. He wrote that it felt like "a universal blow throughout my whole body from head to foot, which seemed within as well as without. . .". He was now using the words *positive* and *negative*, to describe electricity and believed that lightning and electricity were one in the same. He also coined the words *battery, conductor, condenser, electric shock* (firsthand experience), and *electrician*.

In 1748, Franklin retired from the printing profession and devoted his time to science. These were highly productive years and brought him worldwide fame as a scientist. In 1749, Ebenezer Kinnersley, while lecturing on electricity on May 10 in Annapolis, Maryland, first published and demonstrated Franklin's lightning-rod experiments. By 1749, Franklin believed that the world was naturally electrified, and he noted in his journal on November 7 that he thought lightning and electricity were one in the same and needed to prove it.

The following year, he wrote to his friend Collinson and proposed using lightning rods to protect houses in a March 2 letter, but he revised it later to include grounding. On July 29, he devised an experiment involving a device with a pointed rod on its roof, to be erected on a hilltop or a church steeple. During his experiments two days before Christmas, he was severely shocked while electrocuting a turkey.

In April 1751, a collection of Franklin's writings and observations on electricity, titled *Experiments and Observations on Electricity* was published in London. This led him to try his now dangerous but famous experiment in June 1752: to tie a metal key to a kite and to fly it into a thunderstorm on the banks of the Schuylkill. In September, he installed a lightning rod on his house, connecting it to bells that rang when the rod became electrified. Fortunately, his house did not burn down. Confident of his invention, he explained how to perform his kite experiment in the October 19 issue of the *Pennsylvania Gazette*. Franklin also took the time for a side interest: his brother, who was suffering from bladder stones, received some relief when Franklin invented a flexible catheter.

In 1753, he included instructions on how to install lightning rods in his *Poor Richard's Almanak*. Franklin was criticized for his electricity experiments when French physicist Abbé Jean-Antoine Nollet published his "Lettres sur l'Electricité" that same year, disputing Franklin's electrical theories, but Franklin did not respond. Franklin's second set of electrical experiments (*Supplemental Experiments and Observations*) was published in London in March 1753. He was recognized with an honorary Master of Arts degrees from both Harvard (July 25) and Yale (September 12) Universities that same year.

Much of the scientific world did not accept Franklin's views of electricity. In fact, as he writes in his autobiography, a paper he wrote about "the sameness of lightning with electricity" was read and ridiculed at the Royal Academy in London.

After his experiments were repeated in France and the celebrated results were resubmitted to the Royal Academy, there was a different reaction. Franklin writes, "This summary was then printed in their Transactions; and some members of the society in London, particularly the very ingenious Mr. Canton, having verified the experiment of procuring lightning from the clouds by a pointed rod, and acquainting them with the success, they soon made me more than amends for the slight with which they had before treated me."

As a result, the Royal Academy awarded Franklin the Copley medal in 1753, only the second to receive the medal at that time. In 1756, they also made him a Fellow of the Royal Society.

What Franklin did for meteorology turned much of the religious world in a turmoil because it was common belief that lightning came from God. There was a religious and political backlash against Franklin for his invention of the lightning rod. Many religious leaders were furious that Franklin attributed lightning to electricity rather than to God's wrath. Even an earthquake in America in 1755 was blamed on Franklin's rods. The Rev. Thomas Prince, pastor of the Old South Church in Boston, published a sermon and wrote that the frequency of earthquakes may be due to the erection of "iron points invented by the sagacious Mr. Franklin." Moreover, Franklin promoted the use of pointed lightning rods, but Europeans believed that blunt-ended rods were better. The English believed that Americans' favoring of Franklin rods was just another affront to English rule!

In 1760, Franklin's third set of electrical experiments (*New Experiments and Observations on Electricity*) was published in London. In 1769, the fourth edition of *Experiments and Observations on Electricity* was published, and he was elected president of American Philosophical Society, a post he held until his death. The following year, as U.S. postmaster, he collected information about ships sailing between New England and England, discovering and mapping the Gulf Stream for the first time. The following year, Franklin was elected to the Batavian Society of Experimental Science, Rotterdam.

His wife Deborah died in 1774 while he was away in London, and the following year, he left London for Portsmouth to return to America. During the voyage, he tried to understand why sailing from Europe to America took longer than the reverse crossing and measured the temperature of air and water of the sea, showing that the Gulf Stream is warmer than water on either side of it. Only two years later, in 1776, he again went abroad and was sent to France to attempt to gain French assistance in the Revolution, which had already begun.

Always keeping an interest in meteorology, he witnessed two manned balloon flights in 1783 and predicted that balloons would some day be used for military, recreational, and scientific purposes. The idea of daylight saving was first conceived by Franklin in 1784 while he was staying in Paris. At the age of 78, Franklin had just viewed the invention of the oil lamp and decided to write a whimsical piece about lighting in the home. His friend Antoine Alexis-François Cadet de Vaux, editor of the *Journal de Paris*, published his piece as a letter to the editor. Daylight Saving time took hold more than 100 years later in the 20th century; England adopted it in May 1916, and most European countries and America used it during World War I. It was made law in America in 1966 by passage of the federal Uniform Time Act of 1966.

In 1785, on his voyage home from France, he wrote *Maritime Observations*. on the course, velocity, and temperature of the Gulf Stream. One of the last roles he played before his death at age 84, in 1790, was as president of the Pennsylvania Society for Promoting the Abolition of Slavery. On April 9, 1790, he died at home and was buried next to his wife and son Francis in the Christ Church burial ground in Philadelphia.

Franklin was truly one of America's premier scientists and statesmen, contributing to science, politics, and free speech, but of all his accomplishments, one of his best-known and often used quotes relates to climate and weather: "Some are weatherwise; some are otherwise."

⊠   Friday, Jr., Elbert Walter (Joe)
(1938–   )
American
Meteorologist

Elbert (Joe) Friday was born in DeQueen, Arkansas, on July 13, 1939, to Elbert W. Friday, Sr., a communications specialist for the army and air force, and Mary Ward. As the son of a military man, Friday moved frequently and attended grade school in Texas, Georgia, and Tennessee; his high school years followed in Oklahoma, California, and Tennessee. As a youth, Friday's major interests were in science, driving his mother wild with home chemistry experiments and electronic devices. He also worked, setting up pins at a local bowling alley to pay for his 1950 Studebaker while in school. He graduated from Midwest City High School in Midwest City, Oklahoma, in 1957. His nickname Joe came from the 1950s TV series *Dragnet* and whose main character was named Joe Friday. In 1959, he married Karen Ann Hauschild. They had two children.

In 1961, Friday went into the air force as a weather officer in the Air Weather Service. He served as a staff weather officer in the early days of computer applications, providing computer-based weather support to classified operations and worked with the very early versions of the Defense Meteorological Satellite System.

Friday earned three degrees at the University of Oklahoma, receiving a bachelor of science degree in engineering physics (with special distinction) in 1961, a master's degree in meteorology in 1967, and his Ph.D. in meteorology in 1969. He also attended Air Force Command and Staff College, becoming a distinguished graduate in 1972, and the Air War College in 1976, both at Maxwell Air Force Base in Alabama.

Elbert Walter (Joe) Friday, Jr. As director of the National Weather Service, he was responsible for modernizing the agency. *(Courtesy of Elbert Friday, Jr.)*

In 1981, Friday became deputy director of the National Weather Service, and in 1988, he became director, serving until 1997. As deputy, he was responsible for planning the modernization of the Weather Service, and as director was responsible for implementing the plan. This modernization plan included introducing the NEXRAD (Next Generation Radar) Doppler radar, which provides the nation with high-quality radar data and has made possible strides in mesoscale meteorology (subsynoptic or smaller areas of study). The entire modernization has resulted in significantly improved weather warnings and forecasts throughout the nation. The Doppler name comes from Christian Doppler, who discovered in 1853 that the frequency of moving sources of sound shifted according to direction and speed of movement.

Friday is a Fellow of the American Meteorological Society and is its president for 2003. He has been awarded several military awards, including the Air Force Commendation Medal, Meritorious Service Medal, the Bronze Star, and Defense Superior Service Medal. In 1992, he was given the federal government's Presidential Rank Award of Meritorious Executive, and the following year, he received the Distinguished Achievement Award from the University of Oklahoma and the Federal Executive of the Year from the Federal Executive Institute Alumni Association. In 1997, he was awarded the Cleveland Abbe Award by the American Meteorological Society. The following year, he was given the Antonin Stnrad Medal from the Czech Hydrometeorological Institute. In 2000, he received a Group Recognition Award (Outstanding Unit) from the National Academy of Sciences and the 25th Anniversary Award from the National Weather Association. In 2002, he was recognized with the Administrator Special Recognition Award for Service to NOAA.

Friday has published articles in the *Journal of Geophysical Research* and in the *Bulletin of the American Meteorological Society*, and he has given numerous invited presentations before professional organizations, technical and scientific conferences, and international organizations. He has presented numerous testimonies before the House and Senate committees of science, environment, and energy, among others. However, he is most proud of his work in helping to modernize the National Weather Service.

Fritts, Harold Clark
(1928–   )
American
Botanist, Dendrochronologist

Harold Fritts was born on December 17, 1928, in Rochester, New York, to Edwin C. Fritts, a physicist at Eastman Kodak Company, and Ava Washburn Fritts. As a young boy he was interested in natural history and weather and even had a subscription to daily weather maps. Along with his maps, he constructed a weather vane that read out wind directions in his room. Fritts attended Oberlin College in Oberlin, Ohio, from 1948 to 1951, earning a B.A. in botany; from 1951 to 1956, Ohio State University in Columbus, receiving an M.S. in botany in 1953 and Ph.D. in 1956.

During his college years, Fritts was a summer graduate research assistant with the Ohio Agricultural Experiment Station and in 1952 worked in plant genetics at its department of botany and plant pathology. The following summer, Fritts worked at its department of forestry, where he set up study plots of trees infected with oak wilt. In 1954, he took initial measurements on the Cumberland River before dam construction and the following year published his first paper, "A new dendrograph for recording radial changes of a tree," that appeared in the journal *Forest Science*. Developed by his father for use in his master's and Ph.D. research, the dendrograph was an instrument that continuously recorded the radius of a tree. The trace from the instrument shows shrinkage every day due to greater loss of water from the top than enters the root and shows swelling during the night. A gradual increase in the radius represents growth of the ring; analyses of the daily growth rate reveals the daily influence of weather on growth. Several hundred of these instruments were manufactured and sold during his career. In 1955, Fritts married Barbara June Smith; they had two children. He married again in 1982 to Miriam Frances Coulson.

From 1956 to 1960, Fritts was assistant professor of botany at Eastern Illinois University. His father helped develop equipment used in Fritts's doctoral research. In 1960, Fritts was named assistant professor of dendrochronology at the University of Arizona, becoming full professor in 1969; he left the college in 1992 and is now professor emeritus. In 1992, he became an adjunct research associate at the Desert Research Institute in Reno, Nevada, but two years later became owner of Dendro-Power, a dendroecological modeling and consulting service.

Dendrochronology is the study of climate changes and past events by comparing the successive annual growth rings of trees or old timber. Fritts continued to make major contributions on how trees respond to daily climatic factors and how they record that information in ring structure. He developed a method to record the trees' response statistically; using that information, he developed a method to transfer functions to reconstruct climate from past tree rings and to reconstruct spatial arrays of past climate from spatial arrays of tree-ring data. He also developed a biophysical model of tree-ring-structure response to daily weather conditions. This work has laid the groundwork for much current dendroclimatic reconstruction work.

Fritts authored nearly 60 pioneering scientific papers on dendrochronology, including the bible of the field, *Tree Rings and Climate* (1976), one of the most cited books on the subject. In 1965, he was elected a Fellow in the American Association for the Advancement of Science and received a John Simon Guggenheim fellowship in botany in 1968. Fritts was given the Award for Outstanding Achievement in Bioclimatology from the American Meteorological Society in 1982 and, in 1990, received the Award of Appreciation from the Dendrochronological community, in Lund, Sweden.

Fritts pioneered the understanding of the biological relationships and reconstruction of past climate from tree-ring chronologies. He currently is engaged in some scientific writing and is finishing work on the tree-ring model.

G

⊠ **Galilei, Galileo**
(1564–1642)
Italian
Mathematician, Astronomer

Galileo was born on February 15, 1564, in Pisa, Italy, the son of Vincenzo Galilei, a musician and mathematician, and Giulia Ammannati of Pescia. After moving to Florence with his family in 1575, he studied religion, against his father's wishes, and medicine as a young man; he also attended the University of Pisa in 1581, where he studied mathematics and taught for a short time. He took on jobs as an instrument maker, lens grinder, and lecturer.

Before he was 20, he discovered the isochronism of the pendulum, and his treatise in 1588 on the center of gravity gave him a lecturer position at the University of Pisa. However, by 1592, Galileo was a professor of mathematics at the University of Padua. While there for 18 years, he took up experiments in physics particularly with motion and gravity.

Galileo is credited with inventing the first thermometer in 1592. He was also the first to recognize that the expansion of air by heat and its contraction in the cold could be used to measure relative degrees of heating. In actuality, he discovered the thermoscope (there were no graduations). His was an open container of alcohol, with a long glass tube and a sphere at the upper end stuck into the alcohol. As the air temperature changed, the air in the sphere would expand or contract, causing the colored alcohol to move up or down in the tube.

Galileo is also given credit for discovering the telescope, but it is more likely to have been a Dutch lensmaker. Galileo did indeed build and refine a telescope in 1609. Galileo took his refracting telescope and was the first to aim it at the heavens opposite of what they were used for at the time, mostly as a gun-sighting device. In 1609, he observed the moon, the stars, the rings of Jupiter, and four moons of Jupiter (now called the Galilean satellites, but he called them the Medicean stars), the phases of Venus (in 1613), sun spots, and the Sun's rotation.

He was appointed chief mathematician and philosopher by the grand duke of Tuscany and in 1611 was elected a member of the Accademia dei Lincei in Rome and feted by the Jesuit mathematicians of the Roman College. His investigations of the heavens was helping him formulate a defense of Copernicus's Sun-centered universe, not the Earth-centered one currently accepted. He published his findings in a book, *Siderius nuncius (The Starry or Sidereal Messenger)*, which made him famous. His defending later of the theory of Copernicus would get him in trouble with the church.

Galileo Galilei. Galileo is credited with inventing the first thermometer in 1592. It is the development of scientific instruments that helped launch the study of meteorology from folklore to science. *(Courtesy AIP Emilio Segrè Visual Archives)*

Galileo's fame made enemies, and in 1616 he argued his case for the Copernicus theory in front of Cardinal Bellarmine in Rome, who suggested to Galileo that he not write it as fact. In 1624, he received permission to write a book, *Dialogue Concerning the Two Chief World Systems*, explaining the two worldviews—Ptolemy's Earth-centered universe and Copernicus's Sun-centered one. Although the Copernican theory was widely accepted among the intelligentsia, it was not necessarily a challenge to the church because the number of people who understood the theory was limited. It was based on mathematics, and most people of that time did not understand math. However Galileo's book made it possible for anyone to understand the issue.

The book had major impact, and his enemies argued that it was possible to read between the lines that Galileo did indeed favor Copernicus.

He was called in by the pope, Urban VIII, a friend of Galileo's, to stand trial in 1633 for breaking his earlier agreement with the cardinal. In ill health, Galileo was forced to admit to it or face torture and probably death—Giordano Bruno who also supported Copernicus was burned at the stake earlier. Galileo's punishment was house arrest at his home in Arcetri, near Florence.

His banishment did not prevent him from continuing working at his home, making more discoveries and writing his *Discourses on the Two New Sciences,* which was smuggled out of Italy and published in the Netherlands in 1638. He also went blind that year. In 1641, Galileo invited Evangelista TORRICELLI to become his literary assistant and secretary after reading Torricelli's "Trattato del Moto," his thesis on amplifying Galileo's work on projectiles. Antonio Benedetto Castelli, teacher of Torricelli and friend of Galileo, had sent it to him and recommended Torricelli. Galileo posed the problem to Torricelli on how to make a vacuum. He died three months after Torricelli began his work on January 8, 1642, and was buried within the church of Santa Croce in Florence. Torricelli made the discovery in 1644. The Copernican theory was not officially proved until 1838.

Pope John Paul II formally pardoned Galileo from his Inquisition sentence on October 31, 1992, 350 years after Galileo's death. As a testimony to his astronomical work, two lunar features Crater Galilaei and Rima Galilaei are named for him.

⊠ Galton, Sir Francis
(1822–1911)
British
Anthropologist, Explorer

Francis Galton has the distinction of being the half-cousin of another prominent scientist of the 19th century, Charles Darwin. Galton is himself known as the founder of biometry and eugenics.

He was born the youngest of seven children into a wealthy Quaker family in Sparkbrook, near Birmingham, to Samuel Tertius Galton, a banker, and Frances Anne Violetta Darwin, the half-sister of the physician and poet Erasmus Darwin, father of Charles Darwin, who would later influence greatly the mind of Francis.

He was home-schooled by his invalid sister Adele until he was five and was reading at an early age, appearing to have close to instant recall. He later attended King Edward's School in Birmingham between 1836 and 1838 and then became an assistant to the major surgeon in the General Hospital of that city at age 16. He continued his medical education by attending King's College in London and, by 1840, attended Trinity College in Cambridge, although his attention was moving from medicine to mathematics. He never finished his studies due to a nervous breakdown and the stress of taking care of a terminally ill father.

When his father died in 1844, Galton, now wealthy, went exploring, first to Egypt and Syria (1845–46) and then to West South Africa (1850–52) under the approved plan of the Royal Geographic Society. In 1853, he took time to marry Louisa Butler, daughter of the dean of Peterborough; they had no children. She died in 1897.

While in West South Africa, he landed at Walfisch Bay and explored through the interior of what became South-West Africa, now Namibia, to Ovamboland and back, describing the people in a lively and engaging manner. On his return, he wrote his first book *Narrative of an Explorer in Tropical South Africa* in 1853. This resulted in being presented with the medal of the Royal Geographical Society in 1853, and he was made a Fellow of the Royal Society in 1860.

During the 1860s, Galton became interested in meteorology and, in particular, weather patterns, perhaps as a result of his travels. He sent a detail questionnaire to many weather stations in Europe asking for detailed weather conditions that prevailed for a specific month, December

1861. He gathered this information and plotted it on a map in 1862, using the symbols he created, and was able to establish relationships between wind speed and direction and barometric pressure. These maps, which show areas of similar pressure, are the forerunners of today's weather maps. In 1863, he presented his monograph on the subject, *Meteorographica*, to the Royal Society. This may be the first book to deal with modern methods of mapping the weather. He is credited with inventing the term *anticyclone* to describe counterclockwise movements of air that accompanied sudden changes in pressure.

From 1863 to 1867, he served as secretary of the British Association for the Advancement of Science, declining twice the chance to be president. Galton was one of the members of a small departmental committee of the Board of Trade to investigate if the Meteorological Office should be reopened after the suicide of Admiral Robert FITZROY, the former director. Fitzroy committed suicide on April 30, 1865, which historians attribute to a combination of his role in assisting Charles Darwin's career (Fitzroy was antievolution) and criticism by his colleagues of his forecasting abilities.

The result was the formation of a meteorological committee in 1868, of which Galton was a member "for giving storm warnings to seaports, for procuring data for marine charts of weather, and for maintaining a few standard Observatories with self-recording instruments."

This committee was enlarged later as the "Meteorological Council." While a member, Galton improved some of the meteorological instruments being used by the council so data could be recorded and analyzed more efficiently. He also worked on upper-atmosphere issues and ocean currents. After 40 years of working with the council, he resigned in 1905 due to increasing deafness.

By 1865, Galton became keenly interested in genetics and heredity and was influenced by his cousin Charles Darwin's *Origin of Species* in 1859.

In Galton's *Hereditary Genius* (1869), he presented his evidence that talent is an inherited characteristic. In 1872, he took on religion with "Statistical Inquiries into the Efficacy of Prayer." Galton created the study of eugenics, the scientific study of racial improvement and a term he coined, to increase the betterment of humanity through the improvement of inherited characteristics, or as he defined it: "the study of agencies, under social control, that may improve or impair the racial qualities of future generations, either physically or mentally." His thoughts became widely admired on wanting to improve human society.

Galton contributed to other disciplines and authored several books and many papers. In fact, between 1852 and 1910, he published some 450 papers and books in the fields of travel and geography, anthropology, psychology, heredity, anthropometry, statistics, and more. Twenty-three publications alone were on the subject of meteorology.

He became interested in the use of fingerprinting for identification and published *Finger Prints*, the first comprehensive book on the nature of fingerprints and their use in solving crime. He verified the uniqueness and permanence of fingerprints and suggested the first system for classifying them based on grouping the patterns into arches, loops, and whorls.

In 1858, he invented a form of heliostat that sent signals by using the reflection of the Sun and presented it in a paper, "A Hand Heliostat For The Purpose Of Flashing Sun Signals, From On Board Ship Or On Land, In Sunny Climates," to the British Association for the Advancement of Science. He published it in the October 15, 1858, issue of *The Engineer*. A heliostat produces an image of the Sun directly into a laboratory or a room where it can be enlarged for viewing.

As late as 1901, close to 80 years old, he delivered a lecture "On the Possible Improvement of the Human Breed Under Existing Conditions of Law and Sentiment" to the Anthropological Institute and even returned to Egypt for one more visit. He published his autobiography, *Memories of My life*, in 1908 and was knighted the following year. Galton received a number of honors in addition to the ones already cited. He was a member of the Athenaeum Club (1855), received honorary degrees from Oxford (1894) and Cambridge (1895), and was an Honorary Fellow of Trinity College (1902). He was awarded several medals, including the Huxley Medal of the Anthropological Institute (1901), the Darwin-Wallace Medal of the Linnaean Society (1908), and three medals from the Royal Society: the Royal (1886), the Darwin (1902), and the Copley (1910) medals.

Galton will be remembered for being the first to deal with modern methods of mapping weather; terms such as *anticyclone* are now well accepted and used. Galton lived with a grandniece in his later years, and a month short of his 89th birthday, in 1911, his heart gave out during an attack of bronchitis at Grayshott House, Haslemere, in Surrey. He is buried in the family vault at Claverdon, near Warwick, Warwickshire. *Galtonia Candicans*, a white bell-flowered member of the lily family from South Africa, commonly known as the summer hyacinth, was named for Galton in 1888.

⊠   Godfrey, John Stuart
(1940–   )
English
Oceanographer

John Stuart Godfrey was born on November 19, 1940, in Salisbury, England. He was the son of military families on both his father's and mother's sides. Stuart's father Jack threw away a lucrative military retirement in 1950, took his family to Tasmania, and joined the Quakers—a pacifist group; he was passionately antiwar. His mother Elizabeth became a cook on TV in Tasmania and later an elder in the Quaker meeting. Godfrey spent his early education at Summerfields,

Oxford, England, and Hutchins School in Hobart, Australia. He spent three years with the Royal Australian Naval College and then went back to Hutchins.

Godfrey attended the University of Tasmania and received his first bachelor's degree in physics in 1961 and a second in 1962. He moved on to Yale University in New Haven, Connecticut, on a scholarship and received an M.S. in physics in 1963 and a Ph.D. in high-energy physics in 1968. In 1967, he married Nancy Washburn; they have three daughters. In 1971, Godfrey published his first scientific paper, written jointly with A. R. Robinson, in the *Journal of Marine Research*, entitled "The East Australian Current as a Free Inertial Jet." It was Godfrey's first research project and used a simple theory of Professor Allan Robinson's to explore how a strong "western boundary current"—the Gulf Stream is one, and the East Australian Current is another—flows along the east coast of Australia. The model provided an explanation for the strongly eddying character of the East Australian Current.

Godfrey's first job after graduate school—as an oceanographer—was basically the only job of his career. In 1968, he was given a year at Harvard University to "learn what an ocean was," which was much different than his studies of high-energy particles. Since then, his professional career has been with the Commonwealth Scientific & Industrial Research Organization (CSIRO), an Australian government assemblage of research laboratories that dates back to World War I.

During the 1970s and 1980s, Godfrey made two important contributions: the first was published in 1975 exploring how different types of ocean waves combine together to change the state of the Pacific Ocean in an El Niño event. The El Niño–Southern Oscillation Phenomenon is an instability of the "coupled ocean-atmosphere system." Better understanding of El Niño requires deep understanding of how the Pacific (and Indian) oceans respond to and react back on the atmosphere.

John Stuart Godfrey. Godfrey explored how different types of ocean waves combine together to change the state of the Pacific Ocean in an El Niño event. *(Courtesy of John Stuart Godfrey)*

The second was detailed in a paper in 1989 on how the Indonesian Throughflow—a current of some 10 million cubic meters per second that can transport one-quarter of the heat absorbed in the tropical Pacific Ocean to the Indian Ocean—is driven by winds. Godfrey showed that winds along a particular closed path that goes across the Pacific from Peru to New Guinea, down the western Australian coast, across the Pacific again from Tasmania to Chile, and closing along the South American coast should be a prime determinant in the long-term throughflow.

Godfrey has published more than 50 scientific articles, authored or coauthored several books such as *Regional Oceanography: An Introduction* with M. Tomczak Matthias (1994), and written several CSIRO Divisional Reports such as *Tests Of Mixed Layer Schemes And Surface Boundary Conditions In An Ocean General Circulation Model, Using The Imet Flux Data Set.* He received the Sverdrup Medal of the American Meteorological Society in 1999. Eleven years after a divorce from his first marriage in 1981, he married Patricia Moran. He currently is working on the physics of the ocean as related to Earth's climate.

⊠ **Guyot, Arnold Henry (Henri)**
(1807–1884)
Swiss
Geologist

Arnold Guyot was born in Boudevilliers near Neuchâtel, Switzerland, on September 28, 1807. He was educated at Chaux-de-Fonds and then at the college of Neuchâtel, where he was the classmate of Leo Lesquereux, later America's first paleobotanist. In 1825, he moved to Germany and boarded with the parents of botanist Alexander Braun in Carlsruhe, where he met Louis Agassiz. He moved on to Stuttgart, studying there at the gymnasium but returned to Neuchâtel in 1827. He was a very religious man and initially destined for the church, but his interests turned to science. In 1829, after studying under the famous geographer Carl Ritter, he obtained his doctorate on classifying lakes, in Berlin.

For four years, he was employed as a private tutor in Paris, but in the summer of 1838, at Agassiz's request, he visited the glaciers of Switzerland for six weeks. He submitted his findings to the Geological Society of France. He was the first to observe the laminated structure of ice in glaciers, later confirmed by Agassiz and others.

He returned to Neuchâtel and taught as professor of history and physical geography at the Academy of Neuchâtel between 1839 and 1848, along with his colleague Agassiz. However, the grand revolutionary council of Geneva closed the academy in Neuchâtel in 1848, and Guyot fled to America at the urging of his friend Agassiz.

After arriving and settling in Cambridge, Massachusetts, he was invited to deliver a set of lectures at the Lowell Institute in 1848. These lectures, which he delivered in French, titled "Earth and Man," were published in Boston in 1853, after being translated by Cornelius C. Felton of Harvard, and gained for him much acclaim. These lectures also promoted the "new geography" of Ritter (scientific in design) and the concept of "manifest destiny."

Guyot's earliest work however would not be the area of geography, his first love, but rather he was to make an important contribution to meteorology. T. Romeyn Beck, chairman of the New York State Board of Regents of the University of the State of New York and former principal of the Albany Academy where Smithsonian secretary Joseph HENRY went to school, asked his former student Henry for help in cleaning up New York State's system of meteorological observation. Many of the state's teaching academies were working with Henry on his meteorological observation project.

Beck convinced the Board of Regents to hire Guyot to supervise the project. Guyot traveled the state during 1849–50, visiting all the academies and inspecting their equipment as well as tutoring the observers on observation techniques. Appalled at the lack of uniformity in New York's program, he urged Henry to reform the Smithsonian's procedures. Guyot prepared a model for all to follow, and his observations led to the development and implementation of better instruments for the project; by 1851, these instruments were equal to if not superior to those of Europe. Guyot also prepared meteorological tables for Henry that were published in 1852. They allowed anyone to compare instruments made by different makers, as well as correcting tables, and thermometer conversions from many different types of thermometers.

Guyot also convinced the state of Massachusetts that it establish a similar program as in New York to cover the Atlantic Coast. Although New York had established 38 stations by the end of 1850, Massachusetts was swamped with political uncertainty. Guyot proposed that the Smithsonian operate and control a national system in which each state would participate and would help finance. Henry offered Guyot the job of overseeing the Smithsonian Meteorological Project that later developed into the National Weather Bureau, but he declined because he was looking for an academic career. It did not take long.

Although Guyot's early contributions to meteorology are noted, his real interest was in geography. In 1854, he was appointed professor of geology and physical geography at the College of New Jersey (changed to Princeton in 1896) and the following year began what is now the Department of Geological and Geophysical Sciences, giving the first organized teaching in geology at Princeton. He remained the sole teacher of geology until his death in 1884. While there, in 1856 he founded what is now the Princeton Museum of Natural History housed in Nassau Hall (now in Guyot Hall, built in 1909) and continued to contribute specimens to it until his death. He spent much of his time measuring the eastern mountains from New England to North Carolina, including the elevations from Mount Katahdin in Maine to Mount Oglethorpe in Georgia, following what is now the Appalachian Trail.

In 1863, Guyot was one of the 50 founders of the National Academy of Sciences. The society, established in 1863 by an act of Congress, was designed to serve as an official adviser to the federal government on any question of science or technology. Along with Guyot was Joseph Henry who served as second vice president, 1866–68, and president, 1868–78 of the academy.

Guyot surveyed and described the western North Carolina mountain system and reported the information in the manuscript "Notes on the Geography of the Mountain District of Western North Carolina" that was not discovered until 1929. Apparently, some of his research was used, however, in the production of a map he supervised in 1861, with the subsequent production of a western North Carolina map of the area by State Geologist Dr. W.C. Kerr, in 1883. Guyot's original 1861 map served as a basis for developing the boundaries of the Great Smoky Mountains National Park in 1926 and for the naming of the major peaks, including Mount Guyot, named for him.

While surveying, Guyot stayed at various inns or with families. One stay was with a Robert Collins; Guyot named Mount Collins, now in the Great Smoky Mountains National Park, after him. Guyot also named Clingman's Dome and Mount LeConte after a stay at the Montvale Spring Hotel in Tennessee in 1859, named after friends and relatives of the hotel's owner Sterling Lanier. In 1880, he studied the Catskill region of New York State and recognized that Hunter Mountain was the second highest peak in the Catskills (the 4,000-foot Guyot Hill is named for him).

There are three Mount Guyots named in his honor located in the White Mountains of New Hampshire, the North Carolina–Tennessee line in the Great Smoky Mountains (the second highest peak in the state), and the Colorado Rockies. In southeastern Alaska, he has a glacier named for him and the Guyot Crater on the moon. The great flat-topped seamounts or tablemounts—so-called submerged ancient islands—that characterize many parts of the ocean floor were named *guyots* in his honor by Harry H. Hess who, aboard the USS *Cape Johnson* during World War II, utilized the transport's sounding gear, taking thousands of miles of depth soundings, and accidentally led to the discovery of these "ancient islands."

Guyot contributed to America's first national atlas in 1870, and he specialized in his later years in writing textbooks in geography that were used by many. These included an edition in 1876 in the Dakota language called Maka-oyakapi, as well as geographies of specific states. Some of his textbooks include *The Earth and Its Inhabitants. Common School Geography* (1869), *The Earth and Its Inhabitants. Intermediate Geography* (1872), *Physical Geography* (1873), and *Elementary Geography*

For Primary Classes (1879), all part of the Guyot's Geographical Series.

Guyot was a creationist and did not accept organic evolution. He saw the hand of God in the creation of the continents and their structure and promoted this view in his teachings and writings. In 1861, he was a delegate from the U.S. Presbyterian church to the convention of the Evangelical alliance held in Geneva. In 1873, he contributed a paper titled "Cosmogony and the Bible" to a similar New York meeting. In 1884, Guyot published "Creation, or the Biblical Cosmogony in the Light of Modern Science" in the *The New Englander,* and the following year he published *Physical Geography,* a textbook "teaching how the earth is a masterpiece of Divine workmanship, perfect in all its parts and conditions." He also wrote several valuable biographical memoirs on Carl Ritter (1860), James H. Coffin (1875), and Louis Agassiz (1883).

He died after a long illness on February 8, 1884, at Princeton, New Jersey, and is buried at The Princeton Cemetery of the Nassau Presbyterian Church. Recently, a road sign honoring Guyot was placed at Hominy Creek Gap in North Carolina where U.S. 19–23 intersects with Hampton Heights across from Central United Methodist Church. Guyot measured the elevation at sites in western North Carolina from 1856 to 1860, including Hominy Creek Gap, and Mount Guyot, which is located 25 miles northwest of the marker.

Hadley, George
(1685–1768)
British
Physicist, Meteorologist

George Hadley was born in London on February 12, 1685, the son of George Hadley and Katherine FitzJames. He was educated at Pembroke College, Oxford, studied law, and was admitted to the bar in 1709. However, one of his greatest contributions was in meteorology, not law, and had to do with improving Edmund Halley's theory of the general circulation of air.

In 1685, Halley was investigating the oceanic wind systems and currents and published a paper in the *Philosophical Transactions* (16:153–168) titled "An Historical Account of the Trade Winds, and Monsoons, observable in the Seas between and near the Tropicks, with an attempt to assign the Physical cause of the said Winds." In 1735, Hadley improved Halley's observations by giving the big picture of atmospheric circulation when he published his theory for the trade winds, based on conservation of angular momentum, titled "Concerning the Cause of the General Trade-Winds," also published in the *Philosophical Transactions* (39:58–62). Hadley took into account the rotation of the Earth and the displacement of air by tropical heat.

Hadley correctly predicted that the reason air did not flow directly from the north or south near the equator was due to the rotation of the Earth, under forces later developed and explained by CORIOLIS. Because there are high temperatures over the equator and low temperatures over the poles, the spinning Earth breaks the air circulation into three circulatory "cells," or zones.

Surface air rises and flows to the poles from the equator, but at 30 degrees latitude, the air begins to fall or descend, and the returning air flows toward the equator again. The Coriolis effect however deflects the air to the east (right) north of the equator and deflects the air west (left) south of the equator. These surface winds are called the trade winds. This convection cell closest to the equator is named for Hadley because he was the first to propose their existence in explaining the trade winds.

He was elected to Royal Society on February 20, 1735, the same year that he published his paper on the trade winds, and was in charge of recording all official meteorological observations for the society, eventually publishing some of it as the *Account and Abstract of the Meteorological Diaries Communicated for 1729 and 1730*.

He died on June 28, 1768, in Flitton, Bedfordshire. Hadley Crater on Mars was named for him in 1976. His older brother John invented the reflecting telescope and the quandrant (octant), precursor of the sextant, and also had an influence on maritime navigation during the 18th century.

⊠ **Hale, Les**
(1932–   )
American
Electrical Engineer

Les Hale was born in Alamogordo, New Mexico, on June 26, 1932, by accident. His mother was on vacation in the family cabin in Cloudcroft in the mountains of southern New Mexico. This led to a wild ride down the mountains while the doctor drove from El Paso, Texas. Hale is the son of Les Hale, a construction materials salesman and seabee during World War II, and Dorothy Durham Hale.

After earning a bachelor of science degree (1952), a master's degree (1954), and a Ph.D. (1958) in electrical engineering from Carnegie Institute of Technology (now CMU), he was working at Los Alamos Scientific (now National) Laboratory when the Soviet Union launched the space program in the form of the Sputnik satellite in 1957. Americans were surprised, and the country became desperate to do anything to beat the Russians.

At Los Alamos, he did electronic instrumentation (largely classified) and was involved in the invention of an amplifier circuit (largely credited to a person who worked at a company to whom one of Hale's group moved). This circuit was later used as the organizing structure of textbooks on electronic circuitry, which made it possible for Hale to bring relevance to his later teaching.

Los Alamos was winding down on atomic-bomb development and testing and was looking for new things on which to work. Hale was assigned to a small group of "space cadets" to find ways to get Los Alamos vessels into space. Hale's first mission was as part of a group of about a

dozen scientific rocket payloads; because his was the only one that worked, he was assigned to bigger things. One day, he entered NASA's Jet Propulsion Lab (JPL) and left with Los Alamos committing to a mission to Venus. He ended up with about 100 people working on the project; however, it was maneuvered out of his jurisdiction. Finding that he was not enjoying his work because of the perceived need by Los Alamos to concentrate more on money than science, he resigned in 1962. Already well known in "space" circles, he accepted a nine-month appointment as senior visiting fellow in space research at the University of London. While in London, he met Professor Arthur Waynick, who had established the Ionosphere Research Laboratory at Penn State University, and it is here that Hale has spent the rest of his career, to date.

During his first year at Penn State, Hale met Roberta MacMillan. They were married in 1963 and have three children.

Hale's research specialized in the field of atmospheric electricity, essentially lightning electromagnetics, a field he thinks is highly neglected. He cites the neglect of a class of rare but real events of enormous energy, detected on a DOD (Department of Defense) satellite named Blackbeard in assessing the threat to aircraft, pipeline explosions, and the like. Government agencies, such as the NTSB (National Traffic Safety Board), do not consider these events, which could also provide the basis for simple "bistatic" radar to detect new underground facilities in the Middle East, for example.

Hale was principal investigator on more than 100 sounding rockets (a U.S. record) beginning in 1959 and established the electrical "climate" of the mesosphere (about 50 to 85 km) as controlled by mainly invisible aerosol particles. This electrical climate controls the penetration of lightning-related electrical energy to the mesosphere, thus enabling the existence of phenomena such as "red sprites." (These are very large lightning-related optical events that occur in the mesosphere.) Surprisingly, his own Ph.D. thesis, a primarily mathematical treatise finished in 1957 on a seemingly unrelated topic, turned out to be important to the theory of "red sprites."

In 1960, he was the first to confirm the existence of helium in the "exosphere" on an early U.S. "deep space" mission. He has also confirmed the existence of unexpectedly large electric fields in the mesosphere (first observed by the Russians) and proposed a theory of their generation in terms of traveling unipolar ELF "slow tails" from lightning, "polarizing" the Earth's magnetosphere, but it is not yet widely accepted.

He has written dozens of articles and given hundreds of oral presentations. For his last 12 years at Penn State, before retiring in 1992, he held the title "A. Robert Noll Distinguished Professor of Electrical Engineering." His research at Penn State had continuous funding from several sources for more than 30 years, including several years into retirement. He is currently working on a paper describing 50 years of rocket measurements in the mesosphere. He hopes to do a similar retrospective on theoretical aspects of his work at the quadrennial Atmospheric Electricity Symposium in Paris in summer 2003.

Handel, Peter Herwig
(1937–   )
Romanian/American
Physicist

Peter Handel was born on October 16, 1937, in Sibiu (Hermannstadt), Transylvania, Romania. He is the son of Peter Handel, an electrical engineer, and Anna Handel, a high school teacher. As a youngster, Handel conducted physics and chemistry experiments at home and studied physics. He attended the Theoretic Gymnasium of Sinaia, Romania, graduating in 1954, and was a courier in the high Carpathians during highschool summer vacation there. He attended the University of Bucharest, Romania, where he

obtained an M.S. in field theory and one in nuclear physics in 1959. He received a Ph.D. in solid state theory and many-body physics in 1968.

While a student, he published "Free motion of high-speed rockets," which appeared in the *Student Scientific Bulletin of the University of Bucharest* in 1958. This paper included two basic discoveries on relativistic rockets. He also published his first paper related to ball lightning (glowing sphere which drifts horizontally through the air), "Equilibration of a plasma sphere with a single, azimuthally rotating magnetic field and one pre-existing polar magnetic field," in 1959.

Handel has made a number of important discoveries in physics and atmospheric science. Among his discoveries are the quantum l/f (lowfrequency) noise effects. These simple and fundamental effects explain l/f noise as a universal infrared divergence phenomenon. According to this theory, any current, I, with infrared-divergent coupling to a system of infraquanta must exhibit l/f noise with a spectral density of fractional fluctuations of $2a/pfN$ (coherent quantum l/f effect observed in large samples, devices, or systems) or $2aA/fN$ (conventional quantum l/f effect observed in small devices or systems). Here, N is the number of carriers in the sample used for the definition of the current, I. For electric currents, taking the photons as infraquanta, we get a quantum l/f noise contribution with $a=1/137$, the fine structure constant. Finally, $aA=(2a/3p)(Dv/c)2$ is the infrared exponent, essentially the quadratic velocity change in the process considered, in units of the speed of light.

The quantum l/f noise effects are of fundamental nature because they assert that the elementary cross sections and process rates of physics, engineering, or chemistry and biology must fluctuate with a l/f spectral density. Because the error of a l/f-noise limited measurement does not depend on the duration of the experiment, these cross sections and process rates are fundamentally uncertain by an amount of the order of the applicable infrared radiative corrections, unless the number, N, of scattered particles simultaneously present and defining the scattered current is large (which is often not practical). The fluctuations of the elementary cross sections cause fluctuations of kinetic coefficients such as the mobility of current carriers, the conductivity and resistance of the sample or device considered.

The theory has been verified in many ways: l/f noise was found experimentally to be caused by mobility fluctuations, not carrier concentration fluctuations, and the magnitude and temperature dependence was shown to agree with the measurements in submicron metal films and vacuum tubes. Furthermore, l/f fluctuations in the rate of a-radioactive decay, predicted by the theory in 1975 have been verified experimentally, recently also for b-decay. Additional evidence comes from l/f noise in frequency standards and in SQUIDS. Finally, the quantum l/f noise theory has recently been most successful in explaining partition noise in vacuum tubes, and l/f noise in transistors, p–n junctions and infrared radiation detectors. It gives the complicated 1/f noise law in quartz clocks.

This fundamental law of nature was discovered in 1974–75, emerging from Handel's attempts to quantize his turbulence theory to explain 1/f noise in the absence of any turbulence-generating instabilities.

Likewise, Handel contributed to the theory of instabilities and turbulence. The theory has been verified in many ways: l/f noise was found experimentally to be caused by mobility fluctuations, not carrier concentration fluctuations, and the magnitude and temperature dependence was shown to new types of instabilities, such as the thermal and magnetic barrier-instabilities have been found by Handel as a by-product of the work on l/f noise. Handel developed a new magneto-hydrodynamic homogeneous, isotropic, turbulence theory for the plasma of electrons and holes in a semiconductor, as well as for a metal, yielding the first rigorously derived universal classical–physical 1/f spectrum (as was stated by F.N.

Hooge), and the classical analog of his present quantum 1/f effect theory.

Handel also discovered inclusion of spin-orbit effects into the index of refraction formalism for slow neutrons. This problem was a large challenge for Handel because it was considered untreatable by other scientists; the index of refraction depends only on the forward scattering amplitude, and this is zero for spin-orbit scattering! Handel explained for the first time the observed asymmetries in neutron guides.

Handel proposed the many-body theory of static electrification in clouds and thunderclouds. This theory was based on a new type of polarization catastrophe in mixtures of phases containing H_2O aggregates and on the hindered-ferroelectric properties of ice. This theory is in the process of gaining general acceptance. It explains for the first time the ultimate cause of lightning, and it is the first theory to explain in a unified way both terrestrial atmospheric electricity and the electrification processes leading to lightning in Saturn's rings.

Handel also discovered the solution of the "excess-heat" paradox in electrolysis, also known the cold fusion paradox. Handel obtained this solution by introducing a new "thermo-electrochemical effect" and proving that the electrolytic cell works as a thermoelectric heat pump, transporting arbitrarily large amounts of low-grade heat from the environment into the electrolytic cell. It renormalizes electrochemical data and tables, such as Landolt-Börnstein's *Tables of Physical Data*.

Handel solved the excess-heat paradox in the "Patterson cell" and discovered a new heat-pump effect. The thermo-electromechanical effect explains the larger access heat in electro-osmosis and in the Patterson cell. The Patterson cell is an invention of James Patterson, a well-respected inventor who has worked on the possibility of generating excess heat using electrolytic cells.

Patterson proposed the Maser-Soliton theory of ball lightning. This theory, accepted in general today, is related to Pyotr Kapitsa's theory and describes ball lightning as an atmospheric maser of several cubic miles, feeding a localized solitonic field state. This provides a new type of high-frequency discharge, with many industrial applications. A maser, short for microwave amplification by stimulated emission of radiation, is like a laser except that it uses microwaves instead of light.

Handel discovered the universal cause of fundamental 1/f noise spectra: the idempotent (acting as if used only once, even if used multiple times) property of the 1/f spectral form with respect to autoconvolution in 1980 and was reformulated in the practically useful equivalent form of a Universal Sufficient Criterion for the presence of a 1/f spectrum in any homogeneous (h) nonlinear (n) system that exhibits fluctuations. This rigorous mathematical criterion is easy to apply and is stated symbolically h + n = 1/f.

He also proposed two relativistic rocket optimization theorems. These allow for a faster approach to the speed of light and a more reasonable use of fuel in future intergalactic missions.

Handel discovered the 1/Q4-Type of phase noise and improvement of the Leeson formula. This work provided for the first time a correct, simple, and reliable formula for phase noise in resonators, oscillators, and high-tech resonant systems of any type. It includes for the first time Handel's fundamental 1/Q4-mechanism of up-conversion that is present even in the absence of any nonlinearities. It is dominant in highest stability oscillators, resonators, and systems. It includes also fundamental 1/f noise, e.g., quantum 1/f effect found by Handel, adding a 1/Q4-term in Leeson's form.

Handel's contributions are significant in atmospheric science. The many-body theory of static electrification in clouds and thunderclouds (POL-CAT Theory) contributes to the understanding of atmospheric electricity and allows for many practical applications, ranging from controlling weather conditions and protections against lightning, to creating artificial lightning. No artificial lightning, but certainly electric field pulses have been verified reproducibly by Handel with A. L. Chung in his lab. The Maser-Soliton theory of ball lightning introduces a fundamentally new type of high pressure RF discharge at much lower temperatures, with tremendous scientific and practical importance. This was verified through calculations and in part also through experiments at the Kurchatow Institute in Moscow in the program that he has directed since 1992.

Handel has received more than 34 Federal U.S. grants totaling more than $13 million; and grants of $360,000 from Japan, $170,000 from Europe, and $55,000 in private American investment. He has also served as a consultant to many private businesses relating to defense and medical issues and has received a number of patents for his discoveries.

He is the author of more than 190 papers published in refereed journals, in peer-reviewed conference proceedings, or in books edited by other scientists. He was editor of the "Proceedings of the 12th International Conference on Noise in Physical Systems in American Institute of Physics Conference Proceedings #285" and of the "Quantum 1/f Proceedings Series AIP #282, 371, 466." He is a member of the American Physical Society, the American Geophysical Union, a founding member of the European Physical Society, and a senior member of the Institute of Electrical and Electronic Engineers. He has three children.

It is up to his peers to judge whether his most significant contributions in atmospheric electricity are his Pol-Cat theory of atmospheric and cosmic electricity or his Maser-Soliton ball lightning theory. Handel is the originator of new directions in five disciplines of engineering and science and discovered seven new physical effects with applications in science and engineering, allowing for a revolution in high-technology hardware and a revision of quantum notions.

Handel has been a visiting and associate professor at a number of universities, and from 1992 to 2000 he was the director of the International Cooperation on Nonlinear Maser-Soliton Plasma Dynamics at the Kurchatov Atomic Physics Institute in Moscow. This project was designed to investigate the formation of stable plasma solitons (cavitons) in the presence of maserlike excitation at atmospheric pressure. It contained 11 scientists holding Ph.D.s, including three university professors, in three groups. In 1973, he became professor of physics at the University of Missouri at St. Louis, where he continues his research today.

⊠   **Harrison, R. Giles**
(1966–   )
English
Physicist

R. Giles Harrison was born in Stroud, Gloucestershire, United Kingdom, on April 16, 1966, to Barry Harrison, an engineer and one-time meteorologist, and Susan Harrison, a schoolteacher. He attended the local Marling School. As well as science, his school interests included music: he was taught the viola and learned to play piano duets with his sister, Penny.

His early scientific expression was via photography and electronics, and he built scientific instruments with whatever materials he could find or buy locally. He kept weather records from an early age and taught himself and a school friend the technical material to acquire an amateur radio license. (Atmospheric effects on radio propagation were always a strong interest.)

Harrison worked as an analyst in a local chemical factory before he attended Saint Catharine's College at Cambridge University where he received a bachelor in natural sciences in 1988. At Imperial College of the University of London, he received his Ph.D. in atmospheric physics in 1992.

R. Giles Harrison. His work has provided theory and modern instrumental techniques to investigate the weak electrification of the natural atmosphere and its effects on clouds and aerosol particles. *(Courtesy of R. Giles Harrison)*

Harrison's Ph.D. work at Imperial College was supervised jointly by Helen ApSimon and Charles Clement at Harwell Laboratory near Oxford. ApSimon was well known for her work on the transport of radioactivity from the Chernobyl nuclear accident. Harrison had met her when she delivered a talk on the subject to the Cambridge University Meteorological Society, which he was then chairing.

ApSimon and Clement had arranged a project to investigate if the behavior of radioactive aerosols was at all influenced by their electric charges or atmospheric electric fields. The combination of atmospheric science and electrostatics was a field that Harrison had studied in his final undergraduate year, and a clear application made it a compelling topic for him to research. As a result, Clement and Harrison published a Harwell report on "Radioactivity and Atmospheric Electricity" in 1990, and work since then has concentrated on the details of the processes it identified.

Harrison found London to be an expensive and isolating place in which to be a student. However, he used the time to read voraciously the scientific literature in atmospheric electricity and became captivated by the least-known but oldest experimental topic in atmospheric science. He left London to join the world-renowned meteorology department of the University of Reading, Reading Berks, United Kingdom, to put his experimental skills to use in micrometeorology. He subsequently became a lecturer there in 1994 and was able to link the development of instruments and the opportunities for atmospheric experimentation with his long-standing passion for atmospheric electricity. The global aspects of the meteorological work taking place at Reading strongly influenced Harrison's application of atmospheric-electricity work to global climate topics, and he continued to seek new methods of making the basic measurements. Atmospheric electricity is a small field numerically, but Harrison has greatly benefited from the intellectual generosity of many, especially Philip Krider in Arizona and Hannes Tammet in Estonia.

Harrison has made three important contributions. First, working with Charles Clement, he developed a theory for the electrification of radioactive aerosols, published in 1992 and since has been experimentally verified. It showed that radioactive aerosols were not, as previously thought, inclined to discharge themselves by the ionization they produced but could actually acquire large electric charges that modified their behavior.

Secondly, several new approaches have been pioneered for measurements in atmospheric electricity, for monitoring electric fields and air ions, mostly published in the *Review of Scientific Instruments* in the United States. The techniques are at the forefront of what can be achieved with modern electronics and bring laboratory accuracy to the rather variable conditions of the real atmosphere. A recent hybrid instrument, the Programmable Ion Mobility Spectrometer (PIMS) includes a microcontroller to permit combining different operating modes for ion measurements. Its international users include the Danish Space Science Institute, CERN Laboratory (Geneva), and the Pacific Northwest National Laboratory.

Finally, Harrison has uncovered new findings about the atmospheric electrical system and its influence on clouds and aerosols. A recent discovery is the change in the fair-weather atmospheric electric field during the past century. It has been modulated by long-term changes in cosmic rays, caused by variation in solar activity.

Taken together, the work at Reading by Harrison and his colleagues have provided theory and modern instrumental techniques to investigate the weak electrification of the natural atmosphere and its effects on clouds and aerosol particles. A large international consortium of scientists, initially concerned with the a cloud-aerosol project at particle physics facility at the European Organization for Nuclear Research (CERN), has shown that there are major climate questions to which this subject area is central.

Harrison recently chose to pursue archives of historical meteorological data in addition to the more usual laboratory and computer work typical of a physical scientist. As a result, he has recovered early atmospheric electrical data spanning more than a century, from which the long-term atmospheric electrical-field changes became obvious.

The research work he has directed has demonstrated that, rather than being a curious consequence of convective thunderstorms that are traditionally neglected, atmospheric electrification can influence cloud properties. This conclusion is important because thunderstorms occupy only a small fraction of area of the planet, but nonthunderstorm clouds occupy a vast area by comparison; so small electrical influences are therefore important. In short, atmospheric electrical effects on nonthunderstorm clouds are a climate feedback process, urgently requiring further study and quantification.

Harrison has published or presented more than 100 papers and was awarded a visiting scholarship at Oxford University during summer 2001. He has worked industrially on lightning hazards to aircraft and continues his research in the department of meteorology at the University of Reading.

⊠   **Henry, Joseph**
(1797–1878)
American
Physicist

Joseph Henry, the son of teamster William Henry and Ann Alexander, was born in Albany, New York, in 1797, the same year Albany became the official capital of New York State. His father died when Henry was 13, and he spent much of his youth living with his grandmother in the nearby village of Galway. At that age, he was also apprenticed to a watchmaker.

Self-educated, Henry was accepted to the Albany Academy, a private school for boys that opened only seven years after his birth and was to be instrumental in his science career. He attended the academy between 1819 and 1822, while in his 20s, and became a professor of mathematics and natural philosophy (physics) at the academy four years later in 1826.

Besides teaching, Henry pursued an interest in experimental science at the academy that brought him national recognition through his original research on electromagnetics. A year after he began to teach at the academy, Henry presented his first paper on electromagnetism at the Albany Institute on October 10, 1827.

Henry discovered mutual electromagnetic induction—the production of an electric current from a magnetic field—and electromagnetic self-induction, both independently of England's Michael Faraday. Other scientists, such as Christian Oersted, had observed magnetic effects from electric currents, but Henry was the first to wind insulated wires around an iron core to obtain powerful electromagnets.

During the early 1830s, Henry constructed some of the most powerful electromagnets of his time, an oar separator, a prototype telegraph, and the first electric motor. He also is given credit for encouraging Alexander Graham Bell's invention of the telephone. Henry built a 21-pound experimental "Albany magnet" that supported 750 pounds, making it the most powerful magnet ever constructed at the time. His paper describing these experiments and his magnet-winding principle was published in the *American Journal of Science*, a widely read and influential publication, in January 1831.

The following year, Henry became professor of natural philosophy at Princeton University and not only continued his work in electromagnetism but also turned to the study of auroras, lightning, sunspots, ultraviolet light, and molecular cohesion. His interest in meteorology began in Albany while collecting weather data and compiling reports of statewide meteorological observations for the University of the State of New York with his associate T.R. Beck, principal of the academy. An 1825 resolution of the State Board of Regents directed that all academies under their supervision should keep records of the daily fluctuations in temperature, wind, precipitation, and general weather conditions. Ironically, it was Beck who convinced Henry to attend the academy, countering an offer from the Albany Green Street Theater where Henry was pursuing an encouraging acting career. What the theater lost, the world of science gained.

In 1846, Henry was elected secretary of the newly established Smithsonian Institution and guided the institution until his death. He was instrumental in fostering research in a variety of disciplines, including anthropology, archaeology, astronomy, botany, geophysics, meteorology, and zoology. However, one of his first priorities as head of the Smithsonian was to set up "a system of extended meteorological observations for solving the problem of American storms." Basically, Henry began to create the beginning of a national weather service.

By 1849, he had a budget of $1,000 and a network of some 150 volunteer weather observers. Ten years later, the project had more than 600 volunteer observers, including people in Canada, Mexico, Latin America, and the Caribbean region. By 1860, it was taking up 30 percent of the Smithsonian's research and publication budget.

Henry also set up a national network of volunteer meteorological observers, which eventually evolved into the national weather service. The Smithsonian supplied volunteers with necessary instructions, the use of standardized forms, and actual instruments. The volunteers submitted monthly reports that included several observations per day of temperature, barometric pressure, humidity, wind and cloud conditions, and precipitation amounts. Comments were also solicited on events such as thunderstorms, hurricanes, tornadoes, earthquakes, meteors, and auroras.

In 1856, Henry contracted with James H. Coffin, professor of mathematics and natural philosophy at Lafayette College in Easton, Pennsylvania, to interpret the monthly findings. Stacked with as many as a half-million separate observations in a year, Coffin hired several people to conduct the arithmetical calculations and, in 1861, published the first of a two-volume compilation of climatic data and storm observations based on the volunteers' reports for the years 1854–1859.

Henry, whose inventions in the 1830s were instrumental in Samuel Morse's development of the telegraph more than a decade later, saw the benefits of using telegraphy for his weather project. His observations of weather patterns and of storms moving west to east gave him the idea in 1847 that he could use telegraphy to warn the

Joseph Henry. Joseph Henry gave up a promising acting career to teach at the Albany New York Boy's Academy. He discovered mutual electromagnetic induction there, went on to become the first Secretary of the Smithsonian, and created the first network of meteorologists via telegraphs in America. He was the most revered American scientist during the 19th century. *(Courtesy AIP Emilio Segrè Visual Archives, E. Scott Barr Collection)*

northern and eastern part of the country of advancing storms, giving the rise to weather forecasting. He wrote:

> The Citizens of the United States are now scattered over every part of the southern and western portions of North America, and the extended lines of the telegraph will furnish a ready means of warning the more northern and eastern observers to be on the watch from the first appearance of an advancing storm.

By 1857, Henry had a number of telegraph companies transmitting weather data to the Smithsonian; some of them were even supplied with thermometers and barometers. To gain an overview of this information, Henry created a large daily weather map to show weather conditions across the country.

In 1821, William Redfield, an American saddle maker and amateur meteorologist, drew a crude weather map, which may be the first such map that was displayed, beginning in 1856, in the Smithsonian, called the Castle, for the public to view. It became a popular attraction. The map was covered with colored discs, and each disc denoted a different weather condition: white discs meant fair weather, blue ones represented snow, black was rain, and brown was cloudy. Arrows on the discs showed the direction of prevailing winds.

By May 1857, Henry shared the information with the *Washington Evening Star*, which began to publish the daily weather conditions in about two dozen cities, giving rise to the daily weather page that is now commonplace. His map also allowed some forecasting ability, and he planned to predict storm warnings to the East Coast, although the advent of the Civil War prevented the project from continuing. After the war, Henry wrote in his annual report in 1865 that the federal government should establish a national weather service to predict weather conditions. In 1870, Congress put storm and weather predictions in the hands of the U.S. Army's Signal Service, and four years later, Henry convinced the Signal Service to absorb his volunteer observer system.

In 1891, the newly created U.S. Weather Bureau, later the National Weather Service, was created, taking over the weather functions of the Signal Service. Henry's contributions were taken to a new level when Cleveland ABBE was hired by the Signal Service as the government's first weathercaster.

Henry directed the Smithsonian for nearly 32 years but was also president of the American Association for the Advancement of Science (1849–50) and the National Academy of Sciences (1868–1878), among others.

When Henry died, the government closed for his funeral on May 16, 1878, a funeral attended by the president, the vice president, the cabinet, the members of the Supreme Court, Congress, and the senior officers of the army and the navy. Henry was married to Harriet Alexander on May 3, 1830; after his death, Alexander Graham Bell arranged for Harriet to have free phone service out of his appreciation for Henry's early encouragement.

In 1893, the International Congress of Electricians named the international unit of inductance *the henry*, in his honor. A statue of Henry stands under the dome at the Library of Congress. More than a dozen items have been named in his honor. A complete list of these can be found at the Joseph Henry Papers Project website at http://www.si.edu/archives/ihd/jhp/joseph22.html.

⊠ Hertz, Heinrich Rudolph
(1857–1894)
German
Physicist

Heinrich Rudolph Hertz was born in Hamburg, Germany, on February 22, 1857, the son of Gustav F. Hertz, a lawyer and legislator, and Anna Elisabeth Pfefferkorn. As a youngster, he excelled in woodworking and languages, especially Greek and Arabic. He began his education at the University of Munich but transferred to the University of Berlin, becoming an assistant to physicist Hermann von Helmholtz. He received a Doctor of Philosophy, magna cum laude, in 1880.

While Hertz was working with Helmholtz, Helmholtz, like most physicists of the day, was interested in proving or disproving James MAXWELL's theory on electromagnetic waves. In 1879, Helmholtz offered a prize to anyone who could tangibly demonstrate the theory. In 1864,

Maxwell mathematically predicted the existence of electromagnetic waves and suggested that light is part of the electromagnetic spectrum. He recognized that electrical fields and magnetic fields can bond together to form electromagnetic waves, but neither of them move by themselves unless the magnetic field changes; that will then induce the electric field to change, and vice versa, and will advance at the speed of light. It was suggested earlier that electromagnetic waves were similar to sound waves and moved through an unknown medium called luminiferous ether. Scientist Albert A. Michelson, assisted by Edward Williams Morley, shot down this theory in 1881.

In 1883, Hertz was lecturing in physics at the University of Kiel, but two years later (1885), he accepted the position of professor of physics at Karlsruhe Polytechnic in Berlin, staying there only three years. However, while there, in 1886, he married Elizabeth Doll, daughter of a Karlsruhe professor; they had two daughters.

In 1887, while in his physics classroom, Hertz conducted his now famous experiment of generating electric waves by means of the oscillatory discharge of a condenser through a loop that was provided with a spark gap and then detecting them with a similar type of circuit. His homemade condenser was a pair of metal rods, placed end to end, with a small gap for a spark between them. When the rods were given charges of opposite signs strong enough to spark, the current would oscillate back and forth across the gap and along the rods. Hertz proved that the velocity of radio waves was equal to the velocity of light and how to make the electric and magnetic fields separate from the wires and go free as Maxwell's "waves." Further experiments using mirrors, prisms, and metal gratings proved that electromagnetic waves (known also as hertzian waves, or radio waves) had similar properties as light. He demonstrated that these are long, transverse waves that travel at the velocity of light and can be reflected, refracted, and polarized like light. In the process of his investigation, he dis-

Heinrich Rudolph Hertz. His experiments with electromagnetic waves led to the development of the radio, wireless telegraph and radar, the latter two of major importance in meteorology. *(Deutsches Museum, Courtesy AIP Emilio Segrè Visual Archives, Physics Today Collection)*

covered, but did not recognize, the photoelectric effect, the ability to emit light or start a current simply by shining light on a metal surface.

Hertz proved that electricity can be transmitted in electromagnetic waves, which travel at the speed of light and which possess many of the properties of light. His experiments with these electromagnetic waves led to the development of the radio, the wireless telegraph, and radar, the latter two of major importance in meteorology. Ironically, Hertz thought his experiments were useless and only had the effect of proving that Maxwell was right about this theory. Hertz described his experiment in the journal *Annalen der Physik* and later in his first book now considered one of the most important in science, titled *Untersuchungen über die Ausbereitung der elektrischen Kraft* (Investigations on the propagation of electric force).

A teenager named Guglielmo Marchese Marconi read Hertz's journal article and came up with an idea to use radio waves set off by Hertz's spark oscillator for the purpose of sending signals. In 1895, Marconi, using a Hertz oscillator with an antenna and a receiver, successfully transmitted and received a wireless message at his father's home in Bologna. It was a Marconi transmitter that sent an S.O.S. to the Marconi offices in New York and then relayed it to other ships in the vicinity when the *Titanic* went down in 1912. An estimated 800 lives were saved because of it. The rest is history.

In 1889, Hertz succeeded Rudolf Clausius as professor of physics at the University of Bonn. His work there led him to conclude that cathode rays were indeed waves and not particles. In 1890, he published *Electric Waves*. He also remixed Maxwell's theory into a simpler form in 1892 in his publication *Untersuchungen über die Ausbereitung der elektrischen Kraft* (Investigations on the propagation of electric force). His other major works include *Principles of Mechanics* (1894) and *Miscellaneous Papers* (1896).

The unit of frequency that is measured in cycles per second was renamed the hertz in his honor and is commonly abbreviated *Hz*. Most computer users throw the term *megahertz* around without knowing the important contribution its namesake has made in their everyday lives. Unfortunately, Hertz died young of blood poisoning in Bonn on New Year's Day, 1894. He was only 37. One of his eulogies stated that "those who came into personal contact with him were struck by his modesty and charmed by his amiability." The annual IEEE Heinrich Hertz Medal, named in his honor, was established by the Board of Directors in 1987 "for outstanding achievements in Hertzian (radio) waves." The achievements can be theoretical or experimental in

nature. Before his death, he received a number of awards and recognition for his work, including the Royal Society's Rumford Medal in 1890.

⊠   **Holle, Ronald L.**
(1942–)
American
Meteorologist

Ronald Holle was born June 4, 1942, in Fort Wayne, Indiana, to Truman Holle, a wholesale hardware buyer, and Louella Holle. Mixed in with schoolwork at St. Paul's Lutheran Grade School and Concordia Lutheran High Schools in Fort Wayne, Holle had interests in sports, weather, music, and being outdoors. He attended Florida State University and received a bachelor's degree in meteorology in 1964 and an M.S. in the same field two years later.

Holle's research and contributions have focused in three areas: the characteristics of cloud-to-ground lightning as they relate to topography and meteorological factors in time and space over several regions of the United States; the demographics of lightning victims in time and space; and the development of updated guidelines for lightning safety and education. These studies were done shortly after the development of real-time lightning detection systems and are quoted in later studies.

His first published paper in 1966, "Detailed case studies of anomalous winds in jet streams," showed how winds in the jet stream can be much higher than expected by the pressure gradient caused under certain situations. This was the first study to analyze several cases in detail and has been followed by more than 200 more scientific papers and articles by Holle and a number of researchers.

In 1974 Holle married Shirley Feldmann. They have three children. He currently is the senior network applications specialist for Vaisala-GAI (formerly Global Atmospherics, Inc.) in

Tucson, Arizona, and continues working to improve meteorological insight into thunderstorms, understanding of lightning, and the impact of lightning on people and objects.

⊠ Hooke, Robert
(1635–1703)
English
Physicist, Astronomer

Considered one of the greatest scientists of the 17th century and second only to Isaac Newton, Robert Hooke was born in Freshwater, Isle of Wight, on July 18, 1635, the son of John Hooke, a clergyman.

He entered Westminster School in 1648 at the age of 13, and then attended Christ College, Oxford, in 1653 where congregated many of the best English scientists, such as Robert Boyle, Christopher Wren (astronomer), John Wilkins (founder of the Royal Society), and William Petty (cartographer). He never received a bachelor's degree, was nominated for the M.A. by Lord Clarendon, chancellor of the university (1663), and given an M.D. at Doctors' Commons (1691), also by patronage.

Hooke became the assistant to chemist Robert Boyle from 1657 to 1662, and one of his projects was to build an air pump. In 1660, after working with Hooke's air pump for three years, Boyle published the results of his experiments in his paper "New Experiments Physio-Mechanicall, Touching the Spring of the Air and its Effects." Hooke's first publication of his own work came a year after as a small pamphlet on capillary action.

Shortly after, in 1662, with Boyle's backing, he was appointed the first curator of experiments at the newly founded Royal Society of London. The society, also known as The Royal Society of London for Improving Natural Knowledge, was founded on November 28, 1660, after a lecture by Christopher Wren at Gresham College, to discuss the latest developments in science, philosophy, and the arts. The founding fathers, consisting of 12 men, included Wren, Boyle, John Wilkins, Sir Robert Moray, and William Brouncker. This position gave Hooke a unique opportunity to familiarize himself with the latest progress in science.

Part of Hooke's job was to demonstrate and lecture on several experiments at the Royal Society at each weekly meeting. This led him to many observations and inventions in fields that included astronomy, physics, and meteorology. He excelled at this job, and in 1663, Hooke was elected a Fellow of the Society, becoming not just an employee but on equal footing with the other members.

Hooke took advantage of his experience and position. He invented the first reflecting telescope, the spiral spring in watches, an iris diaphragm for telescopes (now used in cameras instead), the universal joint, the first screw-divided quadrant, the first arithmetical machine, a compound microscope, the odometer, a wheel-cutting machine, a hearing aid, a new glass, and carriage improvements. With all this, he is one of the most neglected scientists, due to his argumentative style and the apparent retribution by his enemies such as Newton.

Hooke became a professor of physics at Gresham College in 1665 and stayed there for his entire life. It was also where the Royal Society met until after his death. Hooke also served the society from 1677 to 1683.

The year 1665 is also instrumental for Hooke because he published his major work *Micrographia*, the first treatment on microscopy, in which he demonstrated his remarkable observation powers and ability of microscopic investigation that covered the fields of botany, chemistry, and meteorology. Within this work, he made many acute observations and proposed several theories. Included are many meticulous, hand-made drawings of snow crystals that for the first time revealed the complexity and intricate symmetry of their structure. He also "observ'd such an infinite variety of curiously figur'd *Snow*, that it would be as impossible to draw the Figure and shape of every one of them, as to imitate exactly the curious and Geometrical *Mechanisme* of Nature in any one." This was a suggestion that no two were alike, later proven photographically by Wilson BENTLEY in the 20th century.

Hooke also invented a wind gauge (anemometer) that was similar to Leon Battista ALBERTI's and is often considered the inventor of this wind device. Although Alberti was the first to illustrate one in his book *On the Pleasures of Mathematics* in 1450, there is no evidence that he actually built an anemometer. Hooke was also the first to coin the word *cell* that he attributed to the porous structure of cork, although he failed to realize they were the basic units of life.

In *Micrographia*, he also proposed a wave theory of light and made observations about lunar craters and how they formed. His astronomical contributions, besides creating the first reflecting telescope, also included making the first observation of the rotation of Mars. Hooke discovered the fifth star in Orion, inferred the rotation of Jupiter (1664), and noted one of the earliest examples of a double star. In 1669, he made the earliest attempt to measure the parallax of a fixed star. This led directly to James Bradley's discovery of stellar aberration in 1728.

His contributions were many but in the field of meteorology can be added the inventions of a wheel barometer, a double barometer, and a marine barometer. He also suggested the freezing point of water to be set as the zero point on the thermometer, later refined by Daniel FARHENHEIT and Anders CELSIUS. He also proposed the creation of a weather clock to record barometric pressure, temperature, rainfall, humidity, and wind velocity on a rotating drum.

In 1674, Hooke's *Attempt to Prove the Motion of the Earth* offered a theory of planetary motion in mechanical terms, based on inertia, and a balance between the Earth's centrifugal force and the gravitation attraction of the Sun. Later, in a letter dated January 6, 1680, he wrote to Newton, suggesting that the attraction would vary inversely as the square of the distance between the Sun and the Earth. Hooke anticipated universal gravitation, but he did not have the mathematical ability to formulate his theories into tangible terms. Newton did later, and this led to the feeling by Hooke that Newton stole from him. He did not get along with Christiaan Huygens either about one of his watch inventions.

In 1667, he described how sound could be transmitted using a tightly stretched wire, basically inventing the string telegraph, or as some call it, the string telephone; most school-age youngsters are familiar with this device (tin cans and string). In keeping with this interest, in 1684, he devised a system of telegraphy based on visual signaling. Hooke's interest in flight and air characterics led him to formulate what physicists call Hooke's law: the extension of a spring is proportional to the weight hanging from it. This was published in 1678 in *De Potentia Restitutiva*.

Because of his controversies with competing claims with Christiaan Huygens about the invention of the spring regulator and with Newton, first about optics (1672) and second about the formulation of the inverse square law of gravitation (1686), Hooke fell out of favor in the scientific community. He died in London on March 3, 1703, and was buried in Bishopsgate; however, sometime in the 19th century, his bones were removed. No one knows where he is buried today.

⊠ Howard, Luke
(1772–1864)
English
Chemist, Meteorologist

Luke Howard, called the Godfather of Clouds and the Father of British Meteorology, is an unlikely honoree. He was a successful chemist and pharmacist and the head of his own business in what seems far removed from the halls of meteorology. Yet, as a boy, he was fascinated by clouds, became an amateur meteorologist, and made serious contributions to the field.

Born in London on Red Cross Street on November 28, 1772, he was the first child of Quaker parents Robert and Elizabeth Howard. Howard was sent by his father to a Friends (Quaker) grammar school (Thomas Huntly's School) at Burford, near Oxford, where he studied for seven years (1780–87), especially Latin.

The year 1783 was pivotal in Howard's developing an interest in meteorology. It was a year of atmospheric fireworks. Two volcanic eruptions in Iceland and Japan respectively produced brilliant sunsets and a blanket of ash that covered England. That was topped by earthquakes and a meteor streaking overhead, observed by 10,000 people, including Howard, then 11 years old.

Howard returned to London in 1787 and became a chemist's apprentice and, by 1807, owned his own chemical-manufacturing laboratory. A charter member, on March 23, 1796, of the Askesian Society (Greek for "training"), founded by William Allen, Richard Phillips, and William H. Pepys, he sought with them to promote mutual improvement in matters of science. This club later became The Geological Society of London in 1807. Howard also took the time to marry Mariabell Eliot on December 7, 1796. They had three boys and three girls between 1797 and 1811.

Although the members of the Askesian Society participated in group inhalations of nitrous oxide, it is not known if these experiences gave them any further insight. Howard was a fan of the Swedish botanist Linnaeus, who classified plants and animals using Latin, then the language of educated men and a language with which Howard was well familiar from his childhood schooling. It is no doubt that this systematic thinking of Linnaeus allowed Howard to think about classifying clouds. Before 1800, clouds were considered unlikely to be classified due to their changing shapes and transient nature.

One of the rules of the Askesian Society was that each member had to bring a paper to read and for discussion or pay a fine. In 1802, Howard read his "Essay on Modifications of Clouds," in which he classified four main types: cumulus (Latin for "heap"), stratus (layer), nimbus (rain), and cirrus (curl). He also added intermediate types such as cirro-cumulus, cirro-stratus, and cumulo-stratus. By using Latin names, he made his types universal and widely accepted. This was well received and published the following year in *The Cyclopaedia; or Universal Dictionary of Arts, Sciences and Literature* (by Abraham Rees, originally published in 90 parts between 1802 and 1820) and in *Tilloch's Philosophical Magazine*, vols. XVI, XVII. This was the same year that the naturalist Jean-Baptiste Lamarck published his cloud classification scheme, but it never took hold. Howard's major groups, with a few additions, are still in use today. Even Goethe wrote a poem based on his four cloud types.

In 1818–19, Howard presented his *Seven Lectures on Meteorology* (later published in 1837 as the first textbook on weather). He followed it up in 1821 with the two-volume *The Climate of London* (later three volumes in 1833). On August 3, 1821, he was made a Fellow of the Royal Society for his contributions in meteorology.

In 1847, he published *Barometrographia: Twenty Years' Variation of the Barometer in the Climate Of Britain, Exhibited in Autographic Curves, with the Attendant Winds and Weather, and Copious Notes Illustrative to the Subject*, a booklet on barometric observations made at his home in Tottenham for the years 1815–34. He published as an appendix to this report in 1854, *Papers On Meteorology, Relating Especially To The Climate Of Britain And To The Variations Of The Barometer*.

Howard, a lifelong member of the Society of Friends (Quakers), died in March 1864 at Bruce Grove, Tottenham, Middlesex.

Kalnay, Eugenia Enriqueta
(1942–   )
Argentinian/American
Meteorologist

Eugenia Kalnay was born in Buenos Aires, Argentina, and became a U.S. citizen in 1978. The seventh of eight children, she was the daughter of Jorge Kalnay, an architect in Buenos Aires, and Susana Zwicky, a real estate agent. After attending the very good public schools in Buenos Aires, she went to the University of Buenos Aires, where she received her Licenciatura en Ciencias Meteorologicas in 1965 (a license is a degree in between a bachelor of science degree and a master's degree). Kalnay received a scholarship from the National Weather Service in Argentina and had teaching assistantships from the university. She became a researcher at the university when she graduated. However, when the military took over in 1966 and entered by force into the school of sciences, she resigned, like many of her colleagues.

Kalnay moved to Boston and in 1971 was the first woman to earn a Ph.D. at MIT in meteorology (and the first student in the department to have a baby). Her thesis, "The circulation of the atmosphere of Venus," was mentored by Jule G. CHARNEY. After receiving her Ph.D., she obtained an assistant professorship at the University of Montevideo, Uruguay.

In 1972, she published her first paper, "On the Use of Non-uniform Grids in Finite Difference Equations," in the *Journal of Computational Physics*, followed in 1973 by "Numerical Models of the Circulation of the Atmosphere of Venus" in the *Journal of Atmospheric Science*. She returned to MIT in the fall of 1973, when the military took over the government in Uruguay.

Beginning in 1973, Kalnay was a research associate in the department of meteorology at MIT and in 1977 was promoted to associate professor. In 1979, she moved to NASA Goddard Space Flight Center, where in 1984 she became the head of NASA's global modeling and simulation branch at the Goddard Laboratory for Atmospheres. In 1987, Kalnay became the director of the Environmental Modeling Center (formerly the Development Division) of the National Centers for Environmental Prediction (NCEP), and in 1997, she was named the NCEP deputy for science. In 1999, she became professor and chair of the department of meteorology at the University of Maryland.

Kalnay has extensive research experience and has refereed publications in atmospheric dynamics, general circulation modeling, numerical weather prediction, numerical analysis, the atmosphere of Venus, use of satellite data, atmospheric predictability, data assimilation, and ensemble forecasting (a technique developed that utilizes several forecasts). She has made several important contributions to the study of weather: in her thesis, Kalnay found that the atmosphere of Venus, though very deep, is warmed near the surface by a greenhouse effect and not by a dynamic circulation driven at the cloud level. This discovery explained why the surface of Venus is surprisingly hot. In 1977, she developed fourth-order quadratically conservative finite difference models (used in the NASA fourth-order global model for more than 15 years). Her model at NASA was used and continues to be used for applications of satellite data, among other functions. In 1983, she developed a method (lagged average forecasting, or LAF) for ensemble forecasting. Her method of LAF was the most widely used for ensemble forecasting until 1993 when her more-advanced method of "breeding growing perturbations" became the method for the National Weather Service and other operational centers. Breeding creates very effective perturbations and is now being found to be very useful for data assimilation because the bred perturbations are similar in shape to the forecast errors, which therefore can be removed effectively from the initial conditions.

Kalnay has also made a number of dynamical, numerical, and other contributions in atmospheric predictability, including the most accepted explanation of the origin of closed and open convection. She has developed applications of atmospheric chaos and predictability, physical mechanisms to explain a drought in Texas and Oklahoma in 1998, and improved numerical methods. From 1993 to 1997, she directed the NCEP 50-year Reanalysis Project, which provided researchers, for the first time, with a twice-daily detailed description of the state of the atmosphere throughout the globe. The availability of the reanalysis has resulted into more research papers throughout the world than any previous database. During the decade in which she directed the Environmental Modeling Center of the National Weather Service, Kalnay

oversaw major improvements and new developments. The skill of the operational forecasts doubled during those years. Finally, her new textbook *Atmospheric Modeling, Data Assimilation, and Predictability* (2002) describes, in clear and comprehensive terms, the three central areas of research in numerical weather prediction.

Kalnay has served on many scientific committees, advised a number of M.S., Ph.D., and postdoctoral students, is the author of more than 70 scientific papers, and has served in editorial positions for a number of scientific journals. Among the numerous awards for her work are the NASA Medal for Exceptional Scientific Achievement (1981), Department of Commerce Silver Medal (1990), Department of Commerce Gold Medal awarded to the Development Division (1993), American Meteorological Society Jule G. Charney Award (1995), Senior Executive Service Presidential Rank Award (1996), and the Department of Commerce Gold Medal for the Reanalysis Project (1997). She is a member of the Academia Europaea (1999), and National Academy of Engineering (1996), and a Fellow of the American Meteorological Society (1982); was elected Distinguished University Professor at the University of Maryland (2001); and in 1998–99 was awarded the Robert E. Lowry endowed chair at the University of Oklahoma.

Kalnay married Malise Cooper Dick (1982) and has a son from a previous marriage. Of all her accomplishments, her most satisfying were being director of the NWS Environmental Modeling Center, where during her tenure it was transformed into a leading research and operations center, and directing the Reanalysis Project. Her most important scientific accomplishment was her coinvention with Zoltan Toth of the method of breeding, which is a very simple and effective method to perform perturbations for ensemble forecasting. The availability of ensemble forecasts has become an important tool for human forecasters, who now have a tool that provides them with confidence in the forecasts (if the members of the ensemble are in agreement), or lets them know that the forecasts have lost their skill (when the members of the ensemble give different solutions). To quote one of the NWS regional directors, she put "for the first time, the theory of chaos to practical use." Kalnay continues her work at the University of Maryland where she is currently working on the use of breeding for data assimilation to further increase the skill of the forecasts.

Katsaros, Kristina
(1938–)
Swedish
Meteorologist

Kristina Katsaros. Katsaros was the first woman to earn a Ph.D. in her department at the Atlantic Oceanographic and Meteorological Laboratory, a NOAA Research Laboratory, and the first woman faculty member in that department. She is the first woman in a position as laboratory director in the NOAA research organization. *(Courtesy of Kristina Katsaros)*

Kristina Katsaros was born on July 24, 1938, and raised in Göteborg, Sweden, where she chose a physical-science emphasis in high school at Flickläeroverket i Göteborg, Göteborg, Sweden. She had very supportive parents, who believed that a woman should have as good an education as a man for her independence and security. Her father, G.A. Sander, was an engineer, and her mother, Ester Sander (born Sundstroem), was trained to be a teacher but spent most of her adult life caring for her family full time. After graduation, Katsaros spent a year as an exchange student at the University of Washington in Seattle. Although she was enrolled as a freshman, the school gave her two-thirds of a year credit because of the advanced status of the Swedish high school she had attended. The year 1957–58 was meant to be a break from intense studying: she intended to apply to a chemistry program at the Chalmer's Institute in her hometown when she returned to Sweden; however, life led her down a different path.

In fall 1957, she met Michael A. Katsaros and decided to spend a second year at the University of Washington; that meant looking for something "serious" to study. Because of her interest in chemistry, she checked out the depart-

ment (it was housed in a large, forbidding building) but fortuitously came upon the department of meteorology and climatology, where she walked in and was warmly welcomed by its chair, Dr. Phil Church.

Katsaros signed up for summer school of 1958 and was invited by one of the professors, Konrad J. K. Buettner, to participate in a field program to measure the mountain and valley winds around the beautiful volcano, Mount Rainer. "What a glorious time it was to charge into the mountains at dawn with a group of comrades to set up stations and launch balloons around the clock," she wrote. The team measured the positions of the balloons with theodolites (instruments that find the object in an optical system—like binoculars—and measures the angles of the tube holding the optics). They later calculated the winds up or down slope and to or from the mountain. She was hooked on this kind of fieldwork for life and is forever grateful to her friend and mentor, Buettner, who a few years later became her "thesis father" as they say in his native language, German, when she enrolled in graduate school. Once she had finished her B.S. in meteorology and climatology in 1960 at University of Washington, it was time to take her husband to meet her family in Göeteborg.

They lived there two full years, and she studied some astronomy; worked for a graduate student at the Chalmer's Institute studying the rotation of the arms in our galaxy, the Milky Way, using microwave radiometry; and also participated in the first Swedish pollution study, for which she was supervised by the chief meteorologist at the local airport, Martin William-Olson. Over coffee, they talked and she received his good advice "not to dust away my life." She has since lived up to that advice: she went back to the University of Washington and started a course in atmospheric sciences (the new name of her old department). By 1969, she had her Ph.D. in atmospheric sciences from the University of Washington, and her first child Anthony was then three years old. Her thesis topic was the effect of rain on the so-called cool film on top of the ocean, of interest to the new science of remote sensing from satellites by infrared sensors. A year after the thesis, she had another daughter, Ester. Sadly, her thesis father died in 1970 while at Yale University as a guest professor, something of which he was very proud, but he had greeted her daughter welcome into the world six months earlier. He was very supportive of her dual role as mother and researcher, something exceptional in this attitude for his time. Much credit goes to his artist wife Lucie Buettner who remained Katsaros's friend until her death 30 years later.

Another mentor from her thesis committee, Joost A. Businger, then invited her into his research group, and she began to write her own proposals for funding. The Office of Naval Research funded her work for most of the 1970s when she studied, together with her first graduate student, Dr. W. Timothy Liu, the boundary layer on a convecting water body in laboratory tanks and also in the Arctic Ocean, in a swimming pool, and on Lake Washington (in Seattle). She participated in a field program, the Joint Air-Sea Interaction Program (JASIN) in 1978, in which they had radiation sensors on a ship and she flew in a research aircraft over the North Atlantic Ocean off Scotland. The aircraft carried radiation sensors and so-called infrared radiation thermometers to measure sea surface temperature (SST). Simultaneously, she was developing a field station for measurements of air–sea interaction phenomena, turbulent fluxes of heat and water vapor, and wave generation and their maintenance. Katsaros was also using this station in teaching a summer course each year (since 1976) for international students with financial support from the North Atlantic Treaty Organization (NATO). Her two children, healthy and fine, made it easy for this mother of two to gradually take on full-time and more-demanding work.

In 1978, SEASAT, an experimental satellite, was launched by NASA, and she was able to join one of its science teams. Originally, she thought the new satellite data would contribute to her JASIN research, but her interests quickly expanded with the new fantastic data obtained by microwaves (like radio and radars), so she began to use the new data to study cyclones in midlatitudes, particularly weather fronts. Several graduate students worked with her on the storm analyses and on developing methods for interpreting the satellite microwave signals in terms of atmospheric water vapor content, cloud water content, and precipitation. During the next 25 years, similar instruments were launched in the United States, Europe, Japan, and most recently in India as well.

During the 1980s, Katsaros also continued work on wind and wave research on Lake Washington, which was very much related to the new sensors that infer winds on the ocean surface from the effects of roughness on radar backscatter or emission from the sea. Some of this research in collaboration with European colleagues took place from a research platform in the German Bight in 1979. That Marine Remote Sensing Experiment (MARSEN) was followed by the Humidity Exchange over the Sea Experiment (HEXOS) study in 1984 and 1986, with resulting publications appearing into the mid-1990s. In the 1990s, there was another major field program, the Surface Wave Dynamics Experiment (SWADE) off the East Coast of the United States, where she looked for the effects of varying swell and wind sea on the stress that the wind exerts on the sea surface, which in turns makes the waves grow.

After securing an academic professorship in the atmospheric sciences department at University of Washington in 1983, Katsaros took on more teaching duties, continuing the NATO course summer after summer and visiting Europe for month-long stays in Denmark, the Netherlands, France, Germany, and Portugal. Finally in 1992, she accepted a position in France, where all her experience in microwave satellite and air–sea interaction work came together. She became the first director of the department of oceanography from space at Institut Français de la Recherche et de L'Exploitation de la Mer (IFREMER) in Brest, France. It was a newly formed group of about 20 people, scientists and computer engineers, who were responsible for certain aspects of the data collected by the first European remote Sensing Satellite (ERS 1), which had many instruments that were the offspring (new and improved) of the SEASAT experiment. It was a very exciting and stimulating time period of her life, yet hard work and challenging, especially because it was carried out mostly in French. Katsaros was the only non–French-speaking person, but she had a foundation in that difficult language and received lots of help, especially with formal communications. She could reciprocate by editing the publications and communications with the European

Space Agency. Her slogan in life easily could be "Join Science and Sea the World." Her children were grown and off to college or working, and her husband decided to take early retirement to make this adventure possible.

After five-and-one-half years, Katsaros and family moved back to America, where she received another job offer from another great research organization—as director of the Atlantic Oceanographic and Meteorological Laboratory in Miami, Florida. This research laboratory of the National Oceanic and Atmospheric Administration (NOAA) specializes in climate, hurricane research, and environmental chemistry and biology. She is more of an administrator; the Laboratory has 150 employees, but with good people around her to support the work, she still manages to carry out some of her own research in satellite oceanography/meteorology.

Katsaros's career has followed opportunities on several occasions. She remarks that she could not possibly have set the goal at the start to reach the position where she is today. Much of it has been serendipitous and fortuitous. Colleagues have recognized her with a medal and membership in the National Academy of Engineering. There have been some challenges because she was sometimes threading new territory. Katsaros was the first woman to earn a Ph.D. in her department at the Atlantic Oceanographic and Meteorological Laboratory, a NOAA Research Laboratory, and the first woman faculty member in that department.

She is the first woman in a position as laboratory director in the NOAA research organization. Although all of this attention required adjustments in etiquette for some colleagues, she met universally strong positive support and kindness. Later in her career, Katsaros met a wonderful woman scientist, Joanne SIMPSON, whom she considers a beacon of light from early days of her career because she was aware of Simpson's existence through her published and very eloquently described work in air–sea interaction. All along, her husband, Michael, has been her staunchest supporter, and so have her two children and lately their spouses. There are even two grandchildren, who will learn early about the functioning of rainbows and the growth of ocean waves from their *Mormor* (Swedish for "grandma"). In summing up her life, she says, "It has been a great life, since I was able to so freely follow my bliss, that is, my own inclination and drive. It is easy to share of such happiness and good fortune."

⊠ **King, Patrick**
   (1946–   )
   Canadian
   Meteorologist

Patrick King was born in Wingham, Ontario, to Owen King, a farmer, and Dorothy, who taught in a one-room schoolhouse. The importance of weather there in the Lake Huron Snowbelt and in particular an ice storm in 1958 aroused his interest in meteorology. After the 1958 storm, the ice was so thick that he skated on the roads to go to school, and the ice formed such a thick crust on the snow in the fields that the lighter children actually skated over the fields. When King was young, his father sold the family farm to start a construction business, and often took King to various job sites throughout southwestern Ontario. Almost by osmosis, he became an expert on the geography and physiography of the area. After elementary school in the one-room schoolhouse where his mother taught, he went to Wingham District High School, where he excelled in math and science, winning a number of small scholarships.

His knowledge of the local geography was useful when he began to study the effects of lake breezes on summer severe weather. As he read through the tornado reports, he could see a concentration in certain areas before he even plotted the data. After he plotted the tornado touchdowns on a map, it was quite evident to him that the distribution was affected by lake breezes.

King studied physics and math at Queen's University in Kingston, Ontario, graduating with a B.A. in math and physics in 1968. He married Jean Wong in 1971; they have two daughters. After receiving his bachelor of arts degree, he was hired by the Meteorological Service of Canada and took a nine-month forecaster training course and then worked as a forecaster/briefer for two years at an air force base in New Brunswick and for two more years as a forecaster in Gander, Newfoundland. He then went back to the University of Toronto for an M.S. in atmospheric physics in 1975. He was assigned as a research assistant to B.W. Boville and worked for him for two years, during which time he acted as the forecaster for two field projects: one studied the ozone layer using giant balloons; the other was a cloud-seeding project in northern Ontario.

From 1977 to 1979, King developed refresher courses for weather forecasters. Since 1980, he has researched methods for using satellite data in support of weather forecasting. King's most important discovery concerned the role of lake breezes in triggering severe summer thunderstorms in southern Ontario. Forecasters were aware that thunderstorms in southern Ontario were affected by lake breezes, but the extent and inland penetration were not appreciated. Using tornado climatology, he demonstrated that lake breezes seem to trigger most tornadoes in southern Ontario. He was a principal participant in two field projects studying this effect, ELBOW (The Effects of Lake Breezes On Weather) 1997 and 2001. Most of this work was published in conference proceedings.

Forecasters in Ontario use the ideas that he first suggested in their severe summer forecasting. Ontario is unique in being surrounded by the Great Lakes, and lake breezes from different lakes interact with each other and with cold fronts. Typically, on severe weather days, the surface air arrives from the southwest and almost always interacts with the lake breezes in Ontario. Similar situations occur on the U.S. side of the Great Lakes.

Although his most important contributions are in the effects of lake breezes on severe weather in southern Ontario, King has also worked in rain estimation using satellite imagery and in wind estimation over the oceans using scatterometer data. A high-frequency radar instrument that transmits pulses of energy toward the ocean and measures the backscatter from the ocean surface. It detects wind speed and direction over the oceans by analyzing the backscatter from the small wind-induced ripples on the surface of the water. He is the author of a number of scientific papers and has contributed to several books and encyclopedias. He continues his work as a research meteorologist for the Meteorological Service of Canada.

⊠ Kurihara, Yoshio
(1930–   )
Japanese
Meteorologist

Yoshio Kurihara. Kurihara is most recognized by his peers for his construction of a hurricane-prediction system, the first of its kind, for operational use, leading to remarkable improvement in hurricane track forecast. *(Courtesy of Yoshio Kurihara)*

Yoshio Kurihara, a Japanese meteorologist, was born on October 24, 1930, at Shingishu, Korea, to Jun'ichi Kurihara, an agricultural engineer, and Yoshiko Kurihara. He attended middle school in Tokyo, Japan, graduating in 1948, and graduated from high school, under the old system, the following year. He attended the University of Tokyo in 1949 and received a B.S. in geophysics in 1953. He received a Ph.D. in meteorology from the University of Tokyo in 1962. He married Michiko Ishihara in 1960. They have two children.

From 1953 to 1959, Kurihara was a technical officer at the Japan Meteorological Agency, where he worked on meteorological observation and statistics. In 1956, while at the agency he published his first paper, "Regional numerical weather prediction in a barotropic atmosphere by the method of one-dimensional fourier series with one-dimensional relaxation," in the *Journal of the Meteorological Society of Japan*. After 1959, he was a researcher at Japan Meteorological Agency's Meteorological Research Institute (1959–63, 1965–67) and U.S. NOAA's Geophysical Fluid Dynamics Laboratory (1963–65, 1967–98).

Kurihara has made two important contributions to the study of weather and climate. The first deals with research on processes of structure change of tropical cyclones. This study sheds light on the peculiar behavior of tropical cyclones, such as evolution of storms, their decay after landfall, and interaction with mountains, islands, and oceans. The second contribution is demonstration of the feasibility and merit of the use of a comprehensive three-dimensional dynamical model for operational forecast. The model has made significant reduction in hurricane-forecast errors.

Stimulated by this achievement, similar systems were established in the U. S. Navy, Taiwan, and Korea.

Kurihara has received a number of awards for his work, including the Society Award of the Meteorological Society of Japan (1975) and the Fujiwara Award (1994). The American Meteorological Society presented him the Banner Miller Award in 1984 and again in 1997 and the Jule G. Charney Award in 1996. He received the NOAA Outstanding Scientific Paper Award in 1992 and the Department of Commerce Gold Medal Award in 1993.

Kurihara has published about 80 papers but is most recognized by his peers for his construction of a hurricane-prediction system, the first of its kind, for operational use, leading to remarkable improvement in hurricane track forecast.

L

Langmuir, Irving
(1881–1957)
American
Chemist

Irving Langmuir was born in Brooklyn, New York, on January 31, 1881, the third of four sons of Charles Langmuir and Sadie Comings Langmuir. Langmuir's early education was scattered among various schools in the United States and Paris, and he finally graduated from the Pratt Institute's Manual Training High School in Brooklyn. He attended Columbia University in New York City where he received a bachelor's degree in metallurgical engineering from the university's school of mines in 1903. He attended graduate school in Göttingen University in Germany working with Walther Nernst, a theoretician, inventor, and Nobel laureate, and received his M.A. and Ph.D. in physical chemistry under Nernst in 1906.

Langmuir returned to America and became an instructor in chemistry at the Stevens Institute of Technology, in Hoboken, New Jersey, where he taught until July 1909. He next took a job at the General Electric Company Research Laboratory in Schenectady, New York, where he eventually became associate director of research and development. In 1912, he married Marion Mersereau. They had two children.

Although his studies included chemistry, physics, and engineering, he also became interested in cloud physics. He investigated the properties of adsorbed films and the nature of electric discharges in high vacuum and in certain gases at low pressures, and his research on filaments in gases led directly to the invention of the gas-filled incandescent lamp and the discovery of atomic hydrogen. Langmuir used his discovery of hydrogen in the development of the atomic hydrogen welding process. He formulated a general theory of adsorbed films after observing the very stable, adsorbed, monatomic films on tungsten and platinum filaments and after experiments with oil films on water. He also studied the catalytic properties of such films.

In chemistry, Langmuir's interest in reaction mechanisms led him to study structure and valence, and he contributed to the development of the Lewis theory of shared electrons. In 1927, he invented the term *plasma* for an ionized gas. In 1932, he won the Nobel Prize for his studies on surface chemistry. While at GE, he invented the mercury-condensation vacuum pump, the nitrogen–argon-filled incandescent lamp, and an entire family of high-vacuum radio tubes. He had a total of 63 patents at General Electric. Langmuir also worked with Vincent SCHAEFER, Bernard VONNEGUT, and Duncan BLANCHARD on a number of experiments that included the

Irving Langmuir. Langmuir's discoveries helped shape the establishment of modern radio and television broadcasting, safeguarded the lives of soldiers in war, and provided the framework that allowed his research team to develop a key to possibly control the weather. *(Courtesy of Duncan Blanchard)*

first successful cloud seeding project (making rain) and the development of smoke generators for the WW II effort. Their smoke generator was 400 times more efficient than anything the military had and filled the entire Schoharie Valley within one hour during a demonstration.

Langmuir received many awards and honors, including the Nichols Medal, (1915 and 1920); Hughes Medal (1918); Rumford Medal (1921); Cannizzaro Prize (1925); Perkin Medal (1928); School of Mines Medal (Columbia University, 1929); Chardler Medal (1929); Willard Gibbs Medal (1930); Popular Science Monthly Award (1932); Franklin Medal and Holly Medal (1934); John Scott Award (1937); "Modern Pioneer of

Industry" (1940); Faraday Medal (1944); and the Mascart Medal (1950). He was a foreign member of the Royal Society of London, a Fellow of the American Physical Society, and an honorary member of the British Institute of Metals and the Chemical Society (London). He served as president of the American Chemical Society and as president of the American Association for the Advancement of Science. He received more than a dozen honorary degrees.

Langmuir was an avid outdoorsman and skier and in his early years was associated with the Boy Scout movement, where he organized and served as scoutmaster of one of the first troops in Schenectady, New York. After a heart attack, he died on August 16, 1957, in Falmouth, Massachusetts. In 1975, his son Kenneth Langmuir bequeathed the residue of his estate to the Irving Langmuir Laboratory for Atmospheric Research, where a great deal of lightning research takes place. The bequest supports the laboratory, Langmuir fellowships at New Mexico Institute of Mining and Technology, and an annual research award.

Langmuir's discoveries helped shape the establishment of modern radio and television broadcasting, safeguarded the lives of soldiers in war, and provided the framework that allowed his research team to develop a key to possibly control the weather. Bernard Vonnegut's brother, the writer Kurt Vonnegut, made Langmuir a character "Dr. Felix Hoenikker" in his novel *Cat's Cradle*. Vonnegut claims that the absentminded scientist really did leave a tip for his wife after breakfast one time and abandoned his car in the middle of a traffic jam.

⊠ LeMone, Margaret Anne
(1946–)
American
Meteorologist

Margaret LeMone was born on February 21, 1946, in Columbia, Missouri, to David Vandenberg LeMone, a radiologist who cofounded a cancer hospital, and Margaret Meyer LeMone, a nurse anesthetist. Her interests as a youngster included hiking, playing in the woods, hunting for rocks and fossils, and art, especially sketching cloud formations. In third grade, she became keenly interested in weather when lightning struck her house, exploding part of the roof. She then read everything she could find about weather—especially clouds—and kept weather records through high school; she even conducted some weather forecasting. LeMone attended Grant Elementary School, Jefferson Junior High, and Hickman High School in Columbia and left a year early to attend college at the University of Missouri at Columbia (1963–67), where she obtained a bachelor of arts in mathematics in 1967. While there, however, she worked in the atmospheric science department. She then attended graduate school at the University of Washington (Seattle) and received a Ph.D. in atmospheric sciences in 1972.

She participated in her first research project as a student when she started to plot on map locations where tornadoes were reported in the state of Missouri. The idea was to find out if there were topographic controls. What she and her supervisor, Grant Darkow, found was that there were tornadoes where people were, more tornadoes where newspapers were, and the most tornadoes where a stormchaser lived in the 1950s. Conclusion? You have to see a tornado to report one!

Her discoveries in meteorology deal with clouds and wind. She documented the structure and dynamics of the circulations that create cloud streets (long lines of cumulus clouds visible from space). Some scientists thought they just represented the crests of waves at cloud level; however, the clouds reflect the upwind branch of helical circulations in the subcloud layer. These circulations, called roll vortices (or "rolls" for short) tend to be about 1–2 kilometers deep, counting the clouds. Roll vortices are ubiquitous. Some claim that awareness of the presence of rolls can improve the

representation of the effects of the lower atmosphere in models. The lower atmosphere is the boundary layer—or the lowest 1–2 kilometers in the daytime; this is where it becomes turbulent when you are landing in an airplane.

LeMone also showed that two-dimensional (line) precipitating convection (squall lines, and the like), do not necessarily mix the wind vertically but in fact can increase the vertical shear of the horizontal wind in the direction normal to their axis. Modelers are finding that the effect of line convection on the winds can be important and are starting to figure out how to represent this effect in weather and climate models.

Along with being a member of Phi Beta Kappa and the National Academy of Engineering, LeMone was awarded Woodrow Wilson and National Science Foundation fellowships when she graduated from college. She is a Fellow of both the American Meteorological Society and American Association for the Advancement of Science. She is married to Peter A. Gilman, who studies the dynamics of the solar interior, and they have four children. LeMone has published about 100 articles in referred journals and encyclopedias, has contributed to a high-school earth-science textbook, and is the author of a booklet entitled *The Stories Clouds Tell*, published by the American Meteorological Society. A senior scientist at the National Center for Atmospheric Research, she has been active in educational outreach activities.

LeMone is now working on how the Earth's surface and its heterogeneity affect the distribution of heating and evaporation from the ground.

Leonardo da Vinci
(1452–1519)
Italian
Scientist, Artist

Leonardo da Vinci, one of the greatest minds of all time, was born on April 15, 1452, near the town of Anchiano near Vinci. He was the illegitimate child of a notary, Piero da Vinci, and a peasant woman named Caterina. As a teenager, in 1469, he became an apprentice in one of the best studios in Italy, that of Andrea del Verrocchio, a leading Renaissance master of that period. During this time, da Vinci drew "La valle dell'Arno" (The Arno valley) in 1473 and painted an angel in Verrocchio's "Baptism of Christ" (1475). In 1478, da Vinci became an independent master. Da Vinci is famous for his works of art such as the Mona Lisa, but he is also as famous for his visionary drawings of instruments and machines of the future. He was an artist, scientist, engineer, and architect and one of the first who took detailed observations and experimented in a scientific manner.

Although it is Hypatia, mathematician, astronomer, and Platonic philosopher of Alexandria, who is given credit for inventing the hydrometer in A.D. 400, da Vinci designed two improvements to the instrument that measures the moisture content of the air. In his notebook, Codex Atlanticus, da Vinci designed two hygrometers sometime between 1480 to 1486. One consisted of scales that contained a hygroscopic substance (sponge, cotton wool that absorbed water) in one pan and wax in the other (the wax does not absorb water.) The scales, marked zero on dry days, moved when one of the substances absorbed water from the air. His balance hygrometer represents an improved-upon version of the other design.

Da Vinci has also been given credit for inventing the anemometer, a device that measures wind speed and direction, although that credit is now more rightly given to mathmetician Leon Battista ALBERTI. Da Vinci did design two anemometers, however, between 1483 and 1486. One consisted of a wooden graduated framework and vane that turned by the wind, showing direction. The other was called an anemoscope and also showed the direction of the wind. He also suggested using a clock with it to measure and record the speed of wind.

In his later notebook, the Codex Leicester, one finds the largest assemblage of da Vinci's studies relating to astronomy, meteorology, paleontology, geography, and geology. It reveals that his profound scientific observations far outweigh those of anyone else of his time and also underlines his passion for research and invention. His interest in light and shadow led him to notice how the Earth, the Moon, and the planets all reflect sunlight, for example.

The central topic of the Codex Leicester is the "Body of the Earth" and, in particular, its transformations and movement of water. This study includes a discussion on the light of the moon, the color of the atmosphere, canals and flood control, the effect of the moon on the tides, and modern theories of the formation of continents.

Unfortunately, many of his scientific projects and treatises were never completed because he recorded his technical notes and sketches in numerous notebooks and used mirror script (his writing had to be read in a mirror to be deciphered). It was centuries later that the genius of da Vinci became known. He died at the age of 67 on May 2, 1519, at Cloux, near Amboise, France.

Lilly, Douglas K.
(1929–)
American
Meteorologist

Douglas Lilly was born in San Francisco, California, on June 16, 1929, to Donald Lilly, a construction engineer for Tidewater Oil Company (later Getty), and Dorothy Foster Lilly. Douglas attended school in San Carlos and Redwood City, California, and as a boy had an early interest in science, especially the weather. He began college at Stanford University in 1946, receiving his B.A. in physics in 1950, and then spent three years on active duty in the U.S. Navy (1950–53). After service, he attended graduate school at Florida State University, where he earned an

M.S. in meteorology (1955) and his Ph.D. in meteorology (1959). His dissertation was entitled "On the theory of disturbances in a conditionally unstable atmosphere."

He also found time to court Judith Schuh. They were married in 1955, shortly after she graduated from Florida State University. They have two daughters and a son and six grandchildren.

Lilly spent one year in Munich, Germany, working with Radio Free Europe (1956–57), forecasting weather and winds for a program of flying information leaflets into Eastern Europe by means of constant level balloons. The program ended for political reasons after the Hungarian Revolution. While in graduate school, in 1956, he prepared his first scientific report entitled "A statistical study of some Caribbean wind data," with professor N. LaSeur. The purpose was to evaluate the wind variability in the summer wet season, partly to determine the value of surface winds for analyzing tropical disturbances. Although such disturbances, including hurricanes, often occur, winds in this area are mostly steady. At some stations, topography exerts strong distorting effects; for example, at Port-au-Prince, Haiti, the winds are almost always either from the east or the west, apparently because it lies in a fairly deep east-west valley. Four years later, in 1960, Lilly published his dissertation-related paper on tropical cyclogenesis theory in the *Monthly Weather Review*. This work was among the last serious attempts (ultimately unsuccessful) to apply buoyant convection analysis to tropical cyclone formation. It was followed later by work of K. Ooyama, J. G. CHARNEY, and A. Eliassen, now commonly called convective instability of the second kind (CISK).

During the next 20 years, Lilly made several important advances in the field of meteorology. In 1964, as a senior scientist for the National Center for Atmospheric Research in Boulder, Colorado, he made one of the first successful computer simulations of buoyant convection, that is, the vertical and horizontal motions of heated and cooled air accelerated by gravity. One

of his most widely quoted works (1967) is a method that uses computer simulation to link motion fields, for example, thunderstorm clouds, with very small scale down to centimeters turbulence. This linkage, often called subgrid closure, is necessary to limit realistically the lifetime and energy of the larger scale flows. Also frequently quoted is his work on the structure and driving of marine stratocumulus (1968), the clouds that are so prevalent over and near the California coast and the west coasts of other continents. Lilly's mixed-layer theory remains the foundation for most theoretical descriptions of the marine stratocumulus topped boundary layer (STBL).

In 1969, Lilly used numerical simulation to verify a theory of two-dimensional turbulence developed by Robert Kraichnan in an article in the journal *Physics of Fluids, Supplement*. Before that time, it was thought that turbulence could not exist in two dimensions. Lilly's work on the vertical extent to the tropopause (the boundary between the troposphere and the stratosphere) and above and on the structure of downslope windstorms in Colorado, based on an airplane flight during an intense storm in 1972, is still a primary data source. Peak gusts in excess of 100 miles per hour have been recorded there, but to this day the phenomenon is not completely understood. In 1983 and again in 1986, while professor of meteorology at the University of Oklahoma at Norman, he argued the importance of helicity (a vector product of velocity and rotation rate) in the rotating convective storms that often generate tornadoes. This is one of two mesocyclone paradigms accepted and used today to predict storms and tornadoes.

His work on marine stratus has turned out to have the widest application today because the phenomenon contributes strongly to global climate and probably to human effects on it through aerosol (a gaseous suspension of fine solid or liquid particles) generation. Also, the turbulence closure in one of two or three forms that he proposed is used almost universally in large eddy turbulence simulations in meteorology and engineering. The importance of helicity in convective storms is still somewhat controversial, but helicity is widely used as a forecasting tool.

Lilly has held the George Lynn Cross research professorship at the University of Oklahoma since 1986 and was awarded a distinguished lectureship from 1985–89. From 1992 to 1995, he was the holder of the Robert Lowry endowed chair in meteorology at the University of Oklahoma and is currently an emeritus professor. Since 1997, he has been a distinguished senior scientist at the National Severe Storms Laboratory in Norman, Oklahoma.

Lilly has received a number of awards for his work, including both top research awards of the American Meteorology Society (Jule Charney Award, 1973, and Carl-Gustav Rossby Medal, 1986). He was elected a Fellow of the American Meteorological Society in 1971 and was awarded honorary membership in the society in 2000. Lilly delivered the Symons Memorial Lecture at the Royal Meteorological Society in 1989 and was awarded its top research award, the Symons Gold Medal, in 1993. In 1999, he was elected a member of the National Academy of Science.

Lilly was the leader and principal scientist for a major grant from the National Science Foundation in 1984 for the creation of the Center for Analysis and Prediction of Storms (CAPS), one of the first dozen science and technology centers established by NSF. The center continues to forecast storms.

He has been appointed to numerous directorships and has been a visiting lecturer and scientist to a number of universities and institutes during the last 30 years. He has served on a number of boards and committees and has been a mentor for more than a dozen successful Ph.D. candidates, most of whom are currently employed at research laboratories in the United States and abroad. His more than 90 peer-reviewed publications are found in the majority of scientific journals dealing with atmospheric science.

Lilly has resumed part-time research on marine stratocumulus, and in the summer of 2001, he participated in an aircraft flight program (DYCOMS 2) aimed at evaluating the rate of penetration of upper-level dry air into the cloud layer, which is one of the critical unsolved problems in weather and climate studies.

⊠ Liou, Kuo-Nan
(1944–)
Taiwanese/American
Atmospheric Scientist

Kuo-Nan Liou was born on November 16, 1944, in Taipei, Taiwan, R.O.C. He attended Taiwan University and, in 1965, received a B.S. degree with honors in meteorology. He attended graduate school at New York University where he received both his M.S. (1968) and Ph.D. (1976) in meteorology and atmospheric physics. He has been a naturalized U.S. citizen since 1976.

Liou was a research associate at the Goddard Institute for Space Studies of Columbia University from 1970 to 1972 and an assistant professor (research) at the University of Washington from 1972 to 1974. Appointed associate professor in the department of meteorology, University of Utah, in 1975, he was promoted to professor in 1980. Liou was a visiting scientist at the National Center for Atmospheric Research (summer 1975 and 1976) and NASA Ames Research Center (1980–81); a visiting scholar at Harvard University (1985); and a visiting professor at both the University of California at Los Angeles (1981) and the University of Arizona (1995). He was the director of the Center for Atmospheric and Remote Sounding Studies (CARSS, 1987–97) and adjunct professor of geophysics (1992–97) at the University of Utah, where he served as the chairman of the department of meteorology from 1996 to 1997. He became and remains both honorary professor at Beijing University, China (1990) and adjunct professor of meteorology and physics at the University of Utah (1997). He is presently professor of atmospheric sciences, director of the Institute of Radiation and Remote Sensing (1997), and chair of the department of atmospheric sciences at University of California at Los Angeles (2000).

Liou is the pioneer and leading authority on light scattering by ice crystals and radiative transfer in cirrus clouds. The radiative transfer mechanism drives the atmospheric temperature cycle. Shortly after he received his Ph.D., he constructed the first ice-cloud model taking into account such factors as the orientation of nonspherical particles by using the exact scattering solutions for cylinders as a prototype. He also developed a geometric ray-tracing method for application to light-scattering problems. Liou and his graduate students initiated a geometric-optics/Monte-Carlo approach for computing the scattering, absorption, and polarization properties of hexagonal ice crystals, including the large columns, plates, hollow columns, bullet rosettes, dendrites, double plates, and combinations of those shapes that commonly occur in cirrus clouds. The approach involves the tracing of photons (the quantum of electromagnetic energy, generally regarded as a discrete particle having zero mass, no electric charge, and an indefinitely long lifetime) that undergo localized geometric reflection and refraction (the failure of light to travel in straight lines) and Fraunhofer diffraction (the source of light and the place where you see the light must be relatively far from the obstruction). The fundamental light-scattering results for hexagonal ice crystals with verification from laboratory data have been widely cited and used in conjunction with remote sensing applications and radiation parameterizations (to describe in terms of parameters) involving ice clouds.

Moreover, Liou and his graduate student P. Yang developed a finite-difference-time domain method and a novel geometric-optics/integral-equation method for specific applications to small ice crystals, based on the equivalence theorem for

Kuo-Nan Liou. Liou was the first scientist to demonstrate the theoretical basis and numerical feasibility of inversions leading to the retrieval of atmospheric heating rates. *(Courtesy of Kuo-Nan Liou)*

the exact mapping of the surface electric and magnetic fields determined from the geometric-optics approximation to far field. At this point, Liou has innovated a unified theory of light scattering and absorption by ice crystals of all sizes and shapes, similar to the Mie theory for spherical droplets, which is a breakthrough in the field of atmospheric radiation and cloud physics. A large fraction of clouds in the Earth's atmosphere contain ice particles whose shapes and sizes vary with temperature, supersaturation, and accretion; consequently, Liou's unified theory of light scattering by ice crystals is of fundamental value to the remote sensing of cloud composition and structure from the ground, the air, and space. His theory is articulated in a 1994 review article in *Atmospheric Research* entitled "Light Scattering by Nonspherical Particles: Remote Sensing and Climatic Implications." This article also provides the correct data for parameterization of the radiative properties of ice clouds in mesoscale and climate models. In two recent articles, "Light Scattering and Radiative Transfer in Ice Crystal Clouds: Application to Climate Research" (2000), and "Radiative Transfer in Cirrus Clouds: Light Scattering and Spectral Information" (2001), Liou further demonstrated the relevance and importance of the scattering, absorption, and polarization properties of ice crystals in climate research that involves ubiquitous cirrus clouds, particularly in the tropics.

In the field of radiative transfer, Liou was the first scientist to derive the four-stream solution for radiative transfer in 1994. Further, using the similarity principle, he developed a general delta-function adjustment for the diffraction peak in the scattering-phase function for incorporation in the four-stream solution. It is well suited for applications to the parameterization of radiative flux transfer involving clouds and aerosols in climate and mesoscale models. In 1992, Liou proved that the principles of invariance are "equivalent" to the adding principle of radiative transfer for radiation from above as well as from below. Liou pursued the parameterization of radiative transfer in cloud and aerosol atmospheres for use in GCM (general circulation models) and climate models. He and his graduate student Q. Fu developed a physically based efficient model based on an integration of the delta-four-stream approximation for radiative transfer in nonhomogeneous atmospheres, the correlated k-distribution method for sorting absorption lines, and the scattering and absorption properties of nonspherical ice crystals covering both solar and thermal Infrared spectra. This model, now referred to as Fu and Liou's radiation code, has been used by many scientists and institutions for broadband radiative transfer calculations to study radiative forcing due to the

presence of clouds and aerosols; in fact, this code is currently being used on a routine basis by the CERES (Center for Educational Resources Project)/ARM (Atmospheric Radiation Measurement Program)/GEWEX (Global Energy and Water Cycle Experiment) Experiment (called CAGEX) for the retrieval of radiative fluxes using satellite data. In a recent paper, Liou and his graduate student Y. Gu invented a numerical solution for radiative transfer in three-dimensional nonhomogeneous clouds, based on the unification of a delta diffusion approximation and the correlated k-distribution method, for broadband solar and thermal infrared flux and heating rate calculations, and illustrated its potential for incorporation in GCM and climate models.

In the field of remote sensing, in conjunction with his work on light scattering by ice crystals, Liou was the first to present the theoretical foundation of backscattering depolarization from nonspherical ice crystals. The depolarization approach has become a powerful lidar technique for distinguishing between ice and water clouds, as well as for determining the orientation properties of ice particles. He also proposed a novel technique for the detection of the thickness and composition of cirrus clouds from satellites and demonstrated that cloud-top height and thickness can be determined from a combination of infrared and microwave channels available on the Nimbus satellite sensors. Liou was the first scientist to demonstrate the theoretical basis and numerical feasibility of inversions leading to the retrieval of atmospheric heating rates. He showed that the heating-rate inversion problem can be formulated in terms of the Fredholm equation of the first kind, where the kernel function is a product of the channel transmittance and air density. The required measurements from satellites are a set of radiances in the kernel channel and emergent radiances in the spectral retrieval band, both of which can be measured from a specific angle. This novel technique can be used to derive the tropospheric heating rate, using the rotational band of water vapor directly from satellite measurements. He further demonstrated that if the atmospheric heating rates are known, surface radiative fluxes can be inferred directly from radiation observations at the top of the atmosphere without use of a radiative transfer model.

In collaboration with his longtime associate S. C. Ou, Liou has developed a thermal infrared technique for the detection of high-level clouds and for the retrieval of cirrus-cloud optical depth and mean effective size of ice crystals, based on AVHRR (advanced very high resolution radiometer) channels on board NOAA (National Oceanic and Atmospheric Administration) satellites. For the first time, the fundamental scattering results of hexagonal ice crystals were used in conjunction with parameterizations of radiative transfer and ice microphysics. This technique is presently being considered for incorporation in the NASA EOS/ AM1 satellite cloud algorithm program, as well as in the future National Polar Orbiting Environmental Satellite System (NPOESS) cloud remote sensing program.

In the area of clouds and climate, Liou and his associates developed a one-dimensional, interactive cloud-formation program in connection with a shear-flow convection model to investigate the external radiative forcings on climatic temperature perturbations. This program demonstrates that clouds in the Earth's atmosphere stabilize the surface-temperature increase due to positive radiative forcing, such as an increase in CO_2 concentration. Moreover, recognizing the importance of cloud microphysics processes in climate modeling, Liou developed a cloud-precipitation-climate model to investigate the potential link between the perturbed cloud-particle size distribution produced by greenhouse effects and climate perturbations. The greenhouse effect is the result of atmospheric trace gases, many created by human pollution, that permit incoming solar radiation to reach the surface of the Earth unhindered but restricting the outward flow of infrared radiation. These trace

gases are referred as greenhouse gases and absorb and reradiate this outgoing radiation, storing some of the heat in the atmosphere and producing a net warming of the surface. The process is called the greenhouse effect. If the perturbed mean-particle radii produced by greenhouse effects are less than the climatological mean value, precipitation decreases, leading to increases in cloud liquid water content. Thus, the solar albedo effect outweighs the infrared greenhouse effect, and the Earth's climate may be stabilized. A reduction of the mean droplet radius of about 0.5 μm could cool the atmosphere and offset warming due to CO_2 doubling. In a paper by W. R. Leaitch et al. (1991), a reduction in droplet radii of about 1 μm for eastern North America was observed as a result of anthropogenic (human-caused) pollution. Liou was also the first to note the potential positive feedback associated with particle size in cloud-climate feedback problems when the precipitation factor is taken into consideration.

Liou has served as chairman and coordinator of many committees and projects during the years and in editorial positions for a number of leading journals in atmospheric science; has supervised 24 Ph.D. and 16 M.S. students in atmospheric sciences; and is a Fellow of the Optical Society of America (1983), the American Meteorological Society (1987), American Geophysical Union (1996), and the American Association for the Advancement of Science (2000). He was elected a member of the National Academy of Engineering in 1999 and has received numerous awards, including the creativity award from the National Science Foundation in 1996 and the Jule G. Charney Award from the American Meteorological Society (1998).

Liou has authored and coauthored more than 140 scientific papers in various refereed journals and has presented more than 110 invited and contributed papers at national and international conferences and at academic and research institutions. He unified all the topics associated with the fundamentals of atmospheric radiation into a reference and textbook, *An Introduction to Atmospheric Radiation*, in 1980. Academician K. Kondratyev of the Russian Academy of Sciences wrote, "Liou's monograph is unconditionally useful and timely, successfully filling a gap in publications on radiation of the atmosphere." This text is used by a number of universities for the teaching of atmospheric radiation and is frequently referred to by research scientists in the areas of radiative transfer and light scattering. It has been translated into Russian, Chinese, and Arabic. In 1992, he published a monograph, bridging the fields of radiative transfer, cloud physics, and atmospheric dynamics: *Radiation and Cloud Processes in the Atmosphere: Theory, Observation, and Modeling*. This monograph offers a systematic discussion of the transfer of solar and thermal infrared radiation in the atmosphere and of aspects of cloud processes that are pertinent to radiative transfer.

Liou is currently working in three areas of interest. The first is concerned with the development of remote sensing algorithms for the detection of ubiquitous thin and invisible cirrus clouds and for the retrieval of the optical and microphysical properties of clouds and aerosols with validations utilizing spectral channels that are and will be available from operational research satellites. The second topic deals with the development of an efficient three-dimensional radiative transfer computer program for nonhomogeneous clouds for potential incorporation in cloud and climate models. Finally, he is investigating light scattering and spectroscopic features that involve ice-crystal clouds generated in a laboratory cloud chamber.

M

MacCready, Paul Beattie
(1925–)
American
Physicist

Paul MacCready was born on September 29, 1925, at New Haven, Connecticut, to Dr. Paul Mac-Cready, an ear–nose–throat doctor, and Edith MacCready, a nurse. As a youngster, he was very involved with collecting butterflies and moths and later developed an interest for model airplanes. His interest in flight would become his life's passion. He attended his early schooling at Worthington Hooker in New Haven and went to Hopkins Grammar School for eighth to 12th grades. He set many records for experimental craft, and at age 16, he soloed in powered planes. In World War II, he flew in the U.S. Navy flight-training program.

MacCready went to Yale University, where he received a bachelor's degree in physics in 1947, and attended graduate school at the California Institute of Technology, receiving a master's degree in physics in 1948 and a Ph.D. in aeronautics in 1952. His thesis delineated the spectrum of atmospheric turbulence.

His interest in flight grew to include gliders, and he won the 1948, 1949, and 1953 U.S. National Soaring Championships, pioneered high-altitude wave soaring in the United States and, in 1947, was the first American in 14 years to establish an international soaring record. (The 1999 National Soaring Convention of the Soaring Society of America was dedicated to him.) He represented the United States at contests in Europe four times, becoming International Champion in France in 1956, the first American to achieve this goal.

During his school years, MacCready worked in the physics lab at Yale (1943–44) and then in weather modification programs, 1949–50. In 1950–51, he managed a weather modification program in Arizona. He worked on sailplane development, soaring techniques, and meteorology, and he invented the Speed Ring Airspeed Selector that is used by glider pilots worldwide to select the optimum flight speed between thermals (now commonly called the MacCready speed). In 1957, he married Judith Leonard.

He founded Meteorology Research, Inc., a company that became a leading firm in weather modification and atmospheric science research. He also pioneered the use of small-instrumented aircraft to study storm interiors and performed many of the piloting duties. In 1971, MacCready founded AeroVironment, Inc., a diversified company headquartered in Monrovia, California. The company provides services, developments, and products in the fields of alternative energy, power electronics, and energy efficient vehicles for operation on land and in air and water.

Paul Beattie MacCready. He currently is working on drone aircraft for meteorological long-distance studies, on meteorology of Mars, on solar devices for meteorological measurements, and on all the devices for the physics and measurement of atmospheric properties. *(Courtesy of Paul Beattie MacCready)*

With Department of the Defense (DOD) and then NASA support, his teams moved solar technology into a series of solar-powered stratospheric fliers. The 100-foot *Pathfinder* achieved 71,500 feet in 1997. The 120-feet Pathfinder climbed more than 80,000 feet in 1998. In August 2001, the giant 247-foot *Helios* reached 96,863 feet, more than 2 miles higher than any plane had ever sustained level flight! This flight was a major step toward the aim of "near eternal" (six-month) flight at 55,000 feet. Development is ongoing for the regenerative fuel-cell system that powers night flight, using excess energy stored during daylight. Eventually, such nonpolluting fliers will probe

conditions in the stratosphere, perform surveillance, and serve as 11-mile-high, station-keeping "SkyTower" radio relays for multichannel, widebandwidth telecommunications.

MacCready's discoveries and inventions are particularly useful to atmospheric science, in particular turbulence, and in weather modification, which have been key elements in his career. He is concerned now that the modification of the weather by human effluents is of vital significance for the world to explore.

He has published more than 100 technical reports and articles and given many talks, and he has received a staggering 39 major awards since 1948. MacCready has many professional affiliations, including membership in the National Academy of Engineering, the American Academy of Arts and Sciences, and the American Philosophical Society and has Fellow status in the American Institute of Aeronautics and Astronautics and in the American Meteorological Society. He is also an AMS-certified consulting meteorologist and a member of the AMS Council. MacCready is a Laureate of the Academy of Humanism. For two decades, he has been president of the International Human-Powered Vehicle Association and in 1999 helped create the Dempsey-MacCready One-Hour Distance Prize. He has served on many technical advisory committees and boards of directors for government, industry (public and private corporations), educational institutions, and foundations. At present, MacCready is a director of the Lindbergh Foundation and the Society for Amateur Scientists. He has 15 patents.

MacCready has been awarded five honorary degrees (including one from Yale in 1983) and has made numerous commencement addresses. He has written many popular articles and has authored or coauthored more than 100 formal papers, reports, and journal articles in the fields of aeronautics; soaring and ultralight aircraft; biological flight; drag reduction; surface transportation; wind energy; weather modification; cloud physics; turbulence, diffusion, and wakes; equipment and measurement techniques; and perspectives on technology, efficiency, and global consequences and opportunities.

His most important contribution to meteorology has been in creating Meteorology Research, Inc., and then AeroVironment Inc. He currently is working on drone aircraft for meteorological long-distance studies, on meteorology of Mars, on solar devices for meteorological measurements, and on all the devices for the physics and measurement of atmospheric properties for decades.

⊠ **Mach, Douglas Michael**
(1959–)
American
Physicist

Douglas Mach was born on December 28, 1959, in Denver, Colorado, to Darrell Dean Mach, a civil engineer and mother Dorothy Ann Mach, a geologist. As a youth he was interested in the space program and aircraft and had many a models hanging from his bedroom ceiling. He spent his school days in Preston Elementary School in Rialto, California, Marsteller Middle School in Manassas, Virginia, and high school in Amarillo high school in Amarillo, Texas. While he worked in the construction field—building women's clothing stores—and other jobs, his real goal was to become an astronaut, but his bad eyesight prevented that from happening. Nonetheless, he wanted to work for NASA in some way.

He attended the University of Oklahoma at Norman, receiving a bachelor of science degree and a master's degree in engineering physics in 1982 and 1984, respectively, and a Ph.D. in physics in 1987, the same year he married Aurora Torres. They have three children.

He published his first paper in 1986 with D. R. MacGorman, W. D. Rust, and R. T. Arnold entitled "Site errors and detection efficiency in a magnetic direction-finder network for locating lightning strikes to ground." The paper analyzed data from an early version of the current National Lightning Detection Network (NLDN) to determine the errors in locating lightning strikes to ground (called site errors). These errors can be significant (on the order of 10 degrees or more) and if uncorrected can place the lightning hundreds of miles from its actual strike location.

Working in the area of atmospheric lightning, Mach was one of the first to measure natural, positive return-stroke velocities from positive cloud-to-ground strikes. Lightning typically brings negative charge to ground. A rare form of lightning (10 percent or so) brings positive charge to ground. These positive charge to grounds, or CGs, are often more powerful and cause more damage than the more-common negative CG. So the significance of his paper is that it was the first to document the properties of these rare positive CGs.

Mach has several publications in the area of lightning physics. He continues to conduct his research at the Global Hydrology and Climate Center for the National Space Science and Technology Center, working with NASA, his childhood dream, while at the University of Alabama in Huntsville on a cooperative agreement with NASA. Currently, his projects include looking at the Lightning Launch Commit Criteria (LLCC) for NASA and attempting to improve them, studying lightning optical signals from orbit with two low Earth-orbit lightning satellites and using an unmanned aircraft to study storms in south Florida.

⊠ **Mardiana, Redy**
(1968–)
Indonesian
Electrical Engineer

Redy Mardiana was born on March 4, 1968, in Bandung, Indonesia, to Suherman Mardiana, a civil servant, and mother Sumiyati. He attended high school at SMA Negeri 3 in Bandung, and

Redy Mardiana. Mardiana and his colleagues at the Lightning Research Group of Osaka University have been developing broadband radio interferometers for mapping lightning. *(Courtesy of Redy Mardiana)*

his interests included everything about high-voltage engineering including atmospheric electricity (lightning) and its meteorological aspects.

He attended Institut Teknologi Bandung (ITB) in Bandung and received a bachelor's degree in electrical engineering (1992) and a master's degree in electrical engineering (1997). For his doctoral work, he attended Osaka University in Japan, receiving his Ph.D. in engineering in 2002 under supervision of Professor Zen-Ichiro Kawasaki.

In 1992, Mardiana was a service engineer at a French company (Merlin Gerin Indonesia Co.), and the following year, he moved to the department of electrical engineering at the Institut Teknologi Bandung (ITB) in Bandung, Indonesia, where he currently works as a lecturer. He is also a research fellow in Japan's Osaka University, conducting research in the field of lightning.

Mardiana and his colleagues at the Lightning Research Group of Osaka University have been developing broadband radio interferometers for mapping lightning. The noticeable advantage of broadband interferometry is that it can locate multiple lightning radiation sources, which propagate simultaneously through branching in time and space. This allows detailed observation of the lightning discharge phenomena.

In 2000, he published "A broadband radio interferometer utilizing a sequential triggering technique for locating fast-moving electromagnetic sources emitted from lightning" in the *IEEE Transaction on Instrumentation and Measurement*. This paper addressed the development of broadband interferometers for lightning discharge observation. The system was capable of recording electric field derivative radiation from multiple impulsive events (such as preliminary breakdown, initial and subsequent leaders, cloud discharges) within individual lightning flashes, with submicrosecond time resolution, and of mapping these events in two spatial dimensions and time.

Mardiana has received the Hitachi Scholarship Foundation scholarship (1998–2002) and Bundesministrium fuer Forschungund Technologie (BMFT) fellowship (1992–93). In 1995, he married Effrina Yanti Hamid; they have two sons. He continues to conduct research on lightning phenomena. There are many unknowns to this day regarding lightning discharge, and he feels that his work with broadband interferometry, with some improvements, will help to answer unknowns such as the occurrence of branched lightning channels (the path of lightning bolts).

⊠   Maury, Matthew Fontaine
(1806–1873)
American
Hydrographer, Oceanographer

Matthew Fontaine Maury was born on January 14, 1806, near Fredericksburg in Spotsylvania County,

Virginia, but grew up in Tennessee. He was one of seven children of Richard Maury, a farmer, and Diana Minor. He entered the navy in 1825 at age 19, and the following year, he was assigned to the ship *Vincennes,* which sailed around the world (1826–30). Accompanied by Commodore Jacob Jones's frigate *Brandywine,* the *Vincennes* rounded Cape Horn and cruised in the Pacific until June 1829. From there, the expedition sailed to the Society Islands and the Sandwich (now Hawaiian) chain and reached Macao, China, in 1830. From Macao, he went to the Philippines, then sailed across the Indian Ocean, and arrived at Capetown, South Africa, 56 days later. After a brief stop at St. Helena, he returned to New York on June 8, 1830. This was the first American naval vessel to circumnavigate the globe.

Maury made other voyages as midshipman or astronomer to Europe, the Pacific coast of South America, and the South Seas between 1834 and 1836. This last expedition taught him that reliable navigation charts for much of the Pacific Ocean did not exist. On his return, he married Ann Hull Herndon, resided in Fredericksburg, and had eight children. He did not waste any time writing about his exploits: between 1834 and 1841, he wrote several works on sea navigation and his journeys. However, in 1838 or 1839, Maury suffered a stagecoach accident that left him lame for life and forced him out of active sea duty; yet, he was to become the father of oceanography.

In 1842, Maury was appointed the superintendent of the Depot of Charts and Instruments of the Navy Department in Washington, D.C. The depot is where the navy kept its collection of sea charts, navigational equipment, and other scientific materials.

He began to publish his research on oceanography and meteorology as well as develop charts and sailing directions. Maury set up a system (including races) where sea captains would measure and record a variety of scientific data, and Maury devised a system for organizing and plotting this information on detailed sea charts. This included the first useful maps of the seafloor and currents in the world's oceans, as well as detailed wind charts. Within five years, 25 million reports came in to Maury. Two years later, the depot was renamed the Naval Observatory and became the country's first center for astronomical study. Maury was named the first director of the observatory.

In 1847, he published his first wind-and-current charts for the North Atlantic Ocean. His book *Explanations and Sailing Directions to Accompany the Wind and Current Charts* helped reduce sailing times around the world. Several chapters are devoted to meteorology. A New York–San Francisco route was cut from an average of 187

Matthew Fontaine Maury. Maury's chief contribution to learning, though, was in founding the science of oceanography. His book *Physical Geography of the Sea* (1855), a landmark in American science, encouraged other researchers and explorers to take up oceanography. *(Photo courtesy of John Kasso)*

days to only 144 using his methods. Maury was called the Pathfinder of the Seas.

By 1853, Maury had become recognized around the world and was a representative of the United States at an international congress in Brussels on oceanography and navigation (also known as a conference on meteorology), the first ever of its kind. After that, his system of recording oceanographic data gathered by naval vessels and merchant-marine ships was adopted worldwide. During this same year, Maury discovered a shallow underwater plateau across the Atlantic from Newfoundland to Ireland and proposed it as suitable for holding the wires of a transatlantic cable.

In 1855, he published *The Physical Geography of the Sea*, widely considered the first textbook of modern oceanography and a big seller. Three years later, American industrialist Cyrus Field and the British-based Atlantic Telegraph Company laid the first transatlantic cable along Maury's plateau, creating the first instant communication system between the old and new worlds. Maury received high praise for his work.

Because of his lameness, he did little exploring, but he did organize and garner support for a number of scientific missions. One such was to the North Pacific, led by Lt. Cadwalader Ringgold, who arrived with Lt. Charles Wilkes on the first U.S. exploring expedition in the Pacific in 1841. Ringgold commanded the USS *Porpoise* and led a survey party, mapping the Sacramento River as far as Colusa and also parts of San Francisco Bay. He also supported the 1851 Amazon expedition by William Lewis Herndon.

Maury resigned from the navy on April 20, 1861, three days after Virginia seceded from the Union. His sympathies were with the South. He accepted the position of commander in the Confederate States navy, acquiring war ships and working on harbor defense. He became a spokesman for the South in England. With the war's end, he resided in Mexico where he served (1865–66) under Maximilian as immigration

commissioner in Mexico, attempted to establish colonies of ex-Confederates, and visited England before returning to the United States in 1868 to teach at the Virginia Military Institute as professor of meteorology.

Maury became ill on a lecturing tour, died on February 1, 1873, and was temporarily buried in Lexington. His body was then moved to Hollywood Cemetery in Richmond.

⊠ Maxwell, James Clerk
(1831–1879)
Scottish
Physicist

James Maxwell was born in Edinburgh, Scotland, on June 13, 1831, into a wealthy Scottish family. His father, John Clerk Maxwell, was one of the Clerks of Penicuik, in Midlothian, and his mother, Frances, was the daughter of R. H. Cay, Esq., of North Charlton, Northumberland. His sister Elizabeth died in infancy, and so James was an only child. Maxwell attended the Edinburgh Academy, and at the age of 14, he published his first paper, describing how to draw, using simple mechanical means, such as mathematical curves with a piece of string ("On the description of oval curves, and those having a plurality of foci"). He left the academy at age 16 and entered Edinburgh University. He moved on to Cambridge in 1850 to Peterhouse College but within months moved to Trinity College because it had a better reputation in mathematics.

In 1851, Maxwell took on a tutor, William "Wrangler Maker" Hopkins. Hopkins tutored the likes of Arthur Cayley, Lord (Thomson) Kelvin, George Gabriel Stokes, and Peter Guthrie Tait. Maxwell graduated in 1854 as second wrangler (that is, second highest in the mathematics exam). Edward Routh of Peterhouse, another Hopkins student, beat him, but the two of them were declared joint winners of the highly prestigious Smith Prize. Maxwell graduated in 1854 with a degree in mathematics. He continued graduate work there with a fellowship.

Maxwell turned his attentions to electrical science and consumed Michael Faraday's work. In 1855–56, he showed that mathematically he could express the behavior of electric and magnetic fields and their interrelationships (an oscillating electric charge produces an electromagnetic field).

In November 1856, Maxwell accepted an appointment as chair at Marischal College in Aberdeen, Scotland, after his father died. The following year, Maxwell was determined to win the Cambridge College Adams Prize of 1857; the subject area was *The Motion of Saturn's Rings*. He showed that stability of the rings could only be achieved if the rings consisted of numerous small solid particles, a theory that was proved by spacecraft more than a hundred years later.

In 1860, Maxwell became chair of natural philosophy at King's College in London and stayed there for six years, producing some of his most important experimental work. In 1862, Maxwell showed that light was electromagnetic and calculated that the speed of propagation of an electromagnetic field is about that of the speed of light. In 1866, he formulated (independently of Ludwig Boltzmann), the Maxwell-Boltzmann kinetic theory of gases, showing that temperatures and heat involved only molecular movement. By using statistics, he was able to show that molecules at high temperature have only a high probability of moving toward those at low temperature. In 1871, he became the first Cavendish professor of physics and helped create and design the Cavendish laboratory (opened in 1874). In 1873, his four partial differential equations, known as Maxwell's equations, appeared in *Electricity and Magnetism* and are considered one of the great achievements of 19th-century physics. Between 1874 and 1879, he compiled and edited the papers of Henry Cavendish, *The Electrical Researches of the Honourable Henry Cavendish*, published in 1879. Shortly after, on November 5, 1879, Maxwell died in Cambridge.

James Clerk Maxwell. Maxwell's work is considered to be one of the most influential of the 19th century. His theory of electromagnetic fields led directly to the existence of electromagnetic waves (by Hertz) and was the foundation of today's electric industry, telephony, wireless telegraphy, radio and television. *(Courtesy AIP Emilio Segrè Visual Archives)*

Maxwell's work is considered to be among the most influential of the 19th century. He changed the current thoughts on electromagnetism, later expanded by Heinrich Rudolph HERTZ, Guglielmo Marconi, and Sir Edward Appleton, and was instrumental in introducing the foundation of field theory (the first in electromagnetism), thermodynamics, and the kinetic theory of gases (with Rudolf Clausius). His theory of electromagnetic fields led directly to the existence of electromagnetic waves (by Hertz) and was the foundation of today's electric industry, telephony, wireless telegraphy, radio, and television.

In a 1999 poll taken by the BBC, 100 physicists around the world voted Maxwell as the third-most-important scientist of all time, beaten only by Isaac NEWTON and Albert Einstein.

⊠ McCormick, Michael Patrick
(1940–)
American
Physicist

Michael McCormick was born on November 23, 1940, in Canonsburg, Pennsylvania, to Arthur John McCormick, a maintenance worker for the U.S. Steel Corporation, and Mary Ann Nestor. McCormick attended the Canonsburg Public Schools, and while in school, his interests ranged from various sports (especially wrestling) to trumpet playing to "how things worked."

McCormick attended Washington and Jefferson College in Washington, Pennsylvania, and received his bachelor's degree in 1962 in physics, the same year he married Judy Moyer. They have two children. He completed graduate work at William and Mary College and received physics degrees in 1964 (M.A.) and in 1967 (Ph.D.). In between working as an orderly in Washington Hospital and a summer job at the U.S. Bureau of Mines, he found himself at NASA Langley Research Center in 1967.

Although his first research project was the development of a lidar system for measuring atmospheric aerosols and clouds, he also made several discoveries that have wide implications in meteorology. He has quantified stratospheric aerosols on a global basis; discovered and named Polar Stratospheric Clouds (PSCs); quantified impact of volcanic eruptions on the stratosphere; flew the first Earth-orbiting lidar for atmospheric measurements; quantified Antarctic vortex isolation and subsidence; and developed ozone global climatologies. His research has found that aerosols are important for radiative forcing and climate and that polar stratospheric clouds (PSCs) are important in creation of the ozone hole.

McCormick has received numerous awards for his work, including seven NASA Group or Special Achievement Awards during a period from 1972 to 1988. In 1979, he was presented with the Arthur S. Flemming Award for Outstanding Young People in Federal Service, and two years later, he was awarded an honorary doctor of science degree, conferred by Washington and Jefferson College. He received the Jule G. Charney Award from the American Meteorological Society in 1991 and, in 2000, the Remote Sensing Lecturer Award. Several NASA awards were bestowed on him, including NASA Exceptional Scientific Achievement Medal (1981); H.J.E. Reid Award for Outstanding Paper of 1989 at the NASA Langley Research Center; NASA LaRC Outstanding Paper Award (1990); NASA Outstanding Leadership Medal (1996); and William T. Pecora Award, NASA and Department of the Interior, also in 1996. In 1998, he received the NOAA Environmental Research Laboratories Outstanding Paper Award. Finally, in 2000, he was awarded the NASA Distinguished Public Service Medal.

McCormick's major contributions have been as the principal investigator for the development of satellite remote sensors for measuring on a global basis aerosols, ozone, clouds and water vapor. These include the satellite experiments:

SAM II, SAGE I, SAGE II, SAGE III (passive solar and lunar occultation), and LITE (shuttle-borne lidars). He is continuing his work as an endowed professor at the Center for Atmospheric Sciences at Hampton University in Hampton, Virginia.

⊠ **McIntyre, Michael Edgeworth**
(1941–   )
English
Physicist

Michael McIntyre was born on July 28, 1941, in Sydney, Australia, to Archibald McIntyre, a university research neurophysiologist, and Anne McIntyre, a painter. As a youngster, McIntyre's education was international, starting in nursery school in Sydney, Australia. He then attended P.S. 16 in Yonkers, New York, in 1947, then the Newnham Croft School in Cambridge, England, the following year, followed by George St. Normal School, McAndrew Intermediate School, and King's High, all in Dunedin, New Zealand, from 1949 to 1958. His early childhood interests included classical music (violin, piano, composition), model aircraft, radio, and electronics. At age five, seeing what Beethoven's Eighth Symphony looked like on an oscilloscope made a deep impression. He found occasional work as a local professional musician during his college years, was concertmaster of the New Zealand National Youth Orchestra from 1960 to 1962, and was later to give concerts at the Wigmore Hall and Purcell Room, top professional venues in London, England. He easily could have made a life in music, but his interests turned to science.

He attended the University of Otago in Dunedin, New Zealand, from 1959 to 1962, receiving a B.Sc. (First-class honors) in mathematics in 1963. That same year, he became assistant lecturer in mathematics at Otago. In 1967, he earned his Ph.D. in geophysical fluid dynamics at Cambridge University, his thesis titled "Convection and baroclinic instability in rotating fluids."

McIntyre and his coworkers have made several notable contributions to atmospheric science research, centered around understanding the fluid dynamics of the Earth's atmosphere, with emphasis on the stratosphere, the layer lying between altitudes of about 10 to 50 kilometers. The stratosphere contains the bulk of the ozone shield that protects the Earth from harmful solar ultraviolet radiation. McIntire's research has helped to explain why the strongest humanmade ozone depletion occurs in the Southern Hemisphere in the form of the so-called Antarctic ozone hole, even though the chlorofluorocarbons and other chemicals known to cause it enter the atmosphere mainly in the Northern Hemisphere. Part of the answer is that the chlorofluorocarbon molecules go on epic journeys, circumnavigating the globe and visiting both hemispheres many times in the lower atmosphere before eventually arriving in the stratosphere.

Measured concentrations of chlorofluorocarbons show that they are almost uniformly mixed in the lower atmosphere. Air carrying chlorofluorocarbons is gradually pulled up from the lower atmosphere into the tropical stratosphere and then pushed poleward and back downward into the lower atmosphere by an inexorable, persistent fluid-dynamical process in the stratosphere called gyroscopic pumping.

This is a global-scale pumping action and is called gyroscopic because it depends on the Earth's spin. Complicated, fluctuating fluid motions, which can be compared to giant breaking waves, conspire to push air westward. When air is pushed westward, the Coriolis effect from the Earth's spin tends to deflect it poleward. A movie of the gyroscopic pump in action in the real stratosphere can be viewed at www.atm.damtp.cam.ac.uk/people/mem/.

The instrument that produced the movie in this case uses a set of highly sensitive infrared spectroscopes that are cooled with liquid helium.

Michael Edgeworth McIntyre. McIntyre's research has helped to explain why the strongest humanmade ozone depletion occurs in the Southern Hemisphere in the form of the so-called Antarctic ozone hole. *(Courtesy of Michael Edgeworth McIntyre)*

The global-scale air motion driven by the gyroscopic pumping action upward in the tropics, poleward, and then downward in high latitudes is known to atmospheric scientists as the Brewer-Dobson circulation. Journey through the stratosphere takes several years and exposes the air, the chlorofluorocarbons, and all the other chemicals to hard, solar, ultraviolet radiation along the way. Solar ultraviolet photons high up in the stratosphere are energetic enough to break up even the tightly bound chlorofluorocarbon molecules, prompting a complicated sequence of chemical reactions, some of them forming ozone and others destroying it. The conditions most strongly favoring ozone destruction are found in the Antarctic polar stratosphere in springtime, essentially because those reactions are favored by the extreme cold, but significant ozone destruction has been observed in the Arctic polar stratosphere as well in recent years because global temperature trends are making the Arctic winter stratosphere both colder on average and colder more often, even though less cold than the Antarctic. (The global warming of the lower atmosphere tends to be accompanied by global cooling of the stratosphere.)

McIntyre's work has contributed to the sophisticated mathematical theories used to understand the fluid dynamics involved in all this to describe more precisely the way in which ozone, chlorofluorocarbons, and other chemicals are pumped around in the stratosphere. The mathematical theories go under such technical names as *wave–mean interaction*, *downward control principle*, and *potential-vorticity inversion*. For instance, McIntyre's downward control principle, developed in collaboration with Professor Peter Haynes and others, observes that the pumping action tends to reach mainly downward. More precisely, air pushed westward on a global scale, pumps a global-scale circulation that tends to burrow downward from the altitude of pushing.

Such understanding enables scientists to discern more clearly how the different parts of the atmosphere are linked together: to tell what affects what and how strongly it does so. It also helps with the ongoing effort to improve supercomputer simulations of the atmosphere and other parts of the Earth's life-support system, such as the oceans, the biosphere (the part of the Earth and its atmosphere in which living organisms exist or that is capable of supporting life), and the cryosphere (the portion of the Earth where water is perennially or seasonally frozen as sea ice, snow cover, permafrost, ice sheets, and glaciers). This understanding improves scientists' ability to evaluate the accuracy of computer models; for instance, the downward-control principle made scientists aware that special attention has to be paid to the top boundary of the models, previously considered unimportant.

McIntyre's research interests extend also to other parts of the atmosphere, to the oceans, and to the Sun's interior, where the fluid dynamics are in many ways fundamentally similar. His latest work has led to a breakthrough in understanding how the Sun's differential rotation works. Gyroscopic pumping is again involved, but it is driven this time by the chaotic, turbulent motion in the Sun's outer convection zone and burrowing downward into the region beneath. He has published three papers on this, one of them in collaboration with Professor Douglas Gough, who is best known for his contributions to probing the Sun's interior using sound waves.

McIntyre was elected as a Fellow of the Royal Society (1990), of the American Meteorological Society (1990), and of the American Association for the Advancement of Science (1999). In 1987, he was awarded the Carl-Gustaf Rossby Research Medal, the highest award of the American Meteorological Society, for his work on the fluid dynamics of the stratosphere.

He was a SERC/Engineering and Physical Sciences Research Council senior research fellow from 1992 to 1997, and in 1999, he was awarded the Julius Bartels Medal of the European Geophysical Society. He has published more than 110 scientific papers in a variety of journals. Married to Ruth Hecht and assisting in raising her three children by a previous marriage, McIntyre is currently a professor at the Centre for Atmospheric Science at the Department of Applied Mathematics and Theoretical Physics in Cambridge. He is using his knowledge of music to analyze the "lucidity principles" employed by the best writers and communicators and is writing a book called *Lucidity, Science, and Music or The Two Sides of the Platonic.*

⊠ Milankovitch, Milutin
(1879–1958)
Serbian
Astrophysicist, Engineer

Milutin Milankovitch was born in the rural village of Dalj, Serbia (then Austria-Hungary), on May 28, 1879. He attended the Vienna Institute of Technology and graduated with a doctorate in technical sciences in 1904. He accepted a faculty position in applied mathematics at the University of Belgrade in 1909 shortly after a short stay as an engineer for a construction company, where he built dams, bridges, and other structures. He stayed at the university his entire life.

In 1938, Milankovitch developed one of the prevailing theories relating to Earth motions and long-term climate change. His mathematical theory of climate, based on the seasonal and latitudinal variations of solar radiation (insolation) received by the Earth, was based on previous theories by A. J. Adhemar (1842) and James CROLL (1875). Known as the Milankovitch theory, it states that as the Earth travels through space around the Sun, cyclical variations in the shape of the orbit as it goes around the Sun (orbital eccentricity), changes in the angle that the Earth's axis makes with the plane of the Earth's orbit (obliquity), and precession (the change in the direction of the Earth's axis of rotation) all produce variations in the amount of solar energy that reach the planet and therefore cause the advance and retreat of the polar ice caps and cause glaciation. These orbital motions are known variously as the Milankovitch cycles and the Croll–Milankovitch cycles.

Milankovitch's theory was abandoned when the advent of radiocarbon dating appeared to show that Milankovitch's detailed calculations were in conflict with the timing of the ice ages. It was not until 1976 when a study that examined deep-sea sediment cores using Radiometric (K-Ar) and paleomagnetic dating found Milankovich's theory did indeed corresponded to periods of climate change. Unfortunately, he did not live long enough to see his work accepted. He died on December 12, 1958, in Belgrade.

⊠ **Minnis, Patrick**
(1950–   )
American
Meteorologist

Patrick Minnis was born on October 15, 1950, in Shawnee, Oklahoma, and grew up in Oklahoma City. He is the son of Robert Edward Minnis, Jr., a petroleum geologist, and Kathleen Keefe Johnson, a former social worker. As a youngster, Minnis played many sports (football, basketball, baseball, and track) and enjoyed the outdoors. Like most youngsters, he had a variety of short-lived interests, although when he was nine years old, he spent much of the year making weather observations. He also remembers spending many spring evenings in his basement watching the weather reporter on TV while the city was under a tornado watch. Minnis attended Rosary School, a parochial school in Oklahoma City, through eighth grade, and with a scholarship attended and graduated from Casady School, a prep school in Oklahoma City. Like most youths, he worked odd jobs from gas-station operator at age 14 to telephone sales.

Minnis attended Vanderbilt University in Nashville, Tennessee, and received his bachelor's degree in materials science and metallurgical engineering in 1972. In 1973, he married Deborah K. Mohrman Minnis, whom he met while at Vanderbilt. They have four daughters.

After graduating from Vanderbilt, Minnis began to test products for Ferro Fiberglass in Nashville and after nine months was promoted to quality-control supervisor. A year later, he was placed on the road as a fiberglass salesman. His road job changed his life. As he drove throughout the southeastern United States, he spent many hours watching the sky through his windshield. After he narrowly missed a tornado in Nashville, Tennessee, one evening and another one in Louisville, Tennessee, the next evening during the great tornado outbreak of 1974 (April 3–4), his childhood interest in weather was reawakened. He returned to school the following year in a new field, atmospheric science. He went on to graduate school at Colorado State University in Fort Collins, Colorado, where he received an M.S. in atmospheric science in 1978; he was awarded his Ph.D. in meteorology at the University of Utah, Salt Lake City, in 1991.

With the guidance of his advisor Stephen Cox, he used his earlier experience with X rays in materials science to study atmospheric radiation. He had analyzed potsherds from Indian ruins and clay samples from clay sources in Mexico for an archaeologist who was trying to study trade patterns in pre-Columbian times.

Minnis wrote his first publication in 1976; it appeared as a 1976 Colorado State University technical report and reduced the number of calculations needed to compute the absorption of infrared radiation by water vapor in the atmosphere to facilitate the computation of radiative heating rates in a detailed radiative transfer model. His first journal publication (coauthored) in 1983 in the *Journal of Spacecraft and Rockets* reported the results of a study estimating the errors in the Earth's radiation budget (balance between incoming energy from the Sun and outgoing thermal [longwave] and reflected [shortwave] energy from the Earth) that would result from using various combinations of satellites. It was the first analysis to examine the diurnal variation of clouds and how they would affect the albedo (the fraction of light that is reflected by a body or surface) of the Earth at different times of day. Minnis spent many hours estimating the amount of clouds from satellite photographs and then used the data in a radiative transfer model to simulate what various satellites would measure.

Minnis has made several important contributions in meteorology. He published a three-part article in 1984 that discovered, among other things, the existence of clouds that systematically changed and reformed at relatively predictable times of day over very large areas. His study was the first to use an automated computer algorithm to analyze geostationary satellite data on a large scale. Minnis also developed an empirical model of ocean reflectance patterns that is still being used for satellite analysis. Ocean reflectance contains all the essential information containing the qualitative and quantitative properties of seawater constituents.

His model was the first to objectively quantify the diurnal variations of clouds from infrared and visible satellite imagery and the first to measure the diurnal cycle of reflected and emitted radiation over a significant portion of the Earth.

Minnis's first discovery was important for many reasons. It demonstrated that geostationary digital data could be used scientifically and systematically to study clouds and their optical properties and to estimate the radiation balance. Clouds have a significant effect on the Earth's radiation balance. High, thin cirrus clouds help warm Earth's surface by allowing sunlight to pass through but then trapping heat emitted by the surface. Low, thick cumulus clouds help cool the surface by reflecting incoming sunlight back into space. Prior to that, the satellite images were merely used to study clouds in a qualitative manner or in a very limited scientific manner. The study demonstrated that in addition to surface temperature, the diurnal cycle of clouds should be included in climate models and in remote sensing of the Earth. It also helped pioneer the development of multispectral techniques for cloud remote sensing. Remote sensing is the science of acquiring information about the Earth's surface without actually being in contact with it. This is done by sensing and recording reflected or emitted energy and processing, analyzing, and applying that information.

In 1993, Minnis determined the impact of the 1991 Mount Pinatubo eruption on the Earth's radiation balance and demonstrated that it should cause significant cooling of the Earth. His studies of the Mount Pinatubo event quantified,

for the first time, the actual changes in the Earth's radiation budget resulting from a volcanic eruption. The results, based on satellite data together with other measurements of aerosol loading, were used to model the impact of the eruption on the global surface temperature, leading in turn to accurate determination of the climate impact. The modeling study shows that measurements of radiative changes resulting from volcanic eruptions can be used to predict accurately the changes in the Earth's surface temperature during the following year or two.

Minnis also found in 1993 that for purposes of computing the radiation budget of the Earth and for remote sensing of clouds, it was essential that ice clouds should be modeled, using the optical properties of ice crystals rather than ice spheres. Prior to this study, models assumed that cirrus clouds were composed of ice spheres. In this research, Minnis used the latest theoretical models of ice-crystal optical properties to construct the framework for simulating and remotely sensing cloud properties, using more-realistic hexagonal ice columns. Satellite data using the new models in place of the old spherical particle models showed dramatic differences in the cloud optical depth and height. The new quantities were much more accurate, as determined from aircraft and LIDAR data. LIDAR is an airborne laser system, flown aboard rotary or fixed-wing aircraft, that is used to acquire x, y, and z coordinates of terrain and terrain features that are both human-made and naturally occurring. In atmospheric science, LIDARS are used to profile cloud and aerosol layers as well as to determine humidity profiles in the atmosphere. Minnis's study opened the door for more accurate analyses of satellite data and computations of the radiation budget in cloudy atmospheres. Today, the spherical ice droplet is no longer used to represent ice in clouds in most applications.

In 1998, Minnis discovered the widespread effects of contrails, the streaks of condensed water vapor that airplanes and rockets make in the air. According to Minnis, they can spread into cirrus clouds that cover large areas and are responsible for an increase in cirrus cloudiness over the United States since the beginning of the jet age.

This discovery shows that jet air traffic can have a much larger impact on climate than expected for linear contrails. Minnis demonstrated that contrail-generated clouds can last much longer and can cover up to four times the area predicted by an analysis that assumes that contrails remain linear and narrow. Because of this effect, according to the Minnis study, cirrus cloud cover should be increasing over areas with heavy jet traffic. Initial analyses of cloud observations support this prediction, indicating that cirrus coverage has been increasing over the United States since 1970 despite fluctuations in humidity. Research continues to determine whether contrail-generated cirrus will become a major climate change factor in the future as air traffic continues to increase.

Minnis has been recognized for his pioneering work. In 1985, he received the H.J.E. Reid Award for Most Outstanding Paper at NASA Langley Research Center for "Diurnal variability of regional cloud and clear-sky radiative parameters derived from GOES data." In 1993, he was awarded the NASA Medal for Exceptional Scientific Achievement for outstanding contributions advancing the understanding of clouds and their effect on the Earth's radiation budget and climate. Two years later, NASA again recognized him with the Langley Exceptional Publication Award for "Inference of cirrus cloud properties using satellite-observed visible and infrared radiances." In 1998, the American Meteorological Society honored him with its Henry G. Houghton Award in physical meteorology for outstanding contributions to understanding the effects of clouds and aerosols on the Earth's radiation budget.

Minnis has published more than 100 papers and presented more than 200 presentations and papers in his career, but more important, he has been instrumental in expanding the scientific study of clouds using geostationary satellite data.

⊠ **Mintz, Yale**
(1916–1991)
American
Meteorologist

Yale Mintz is widely known as a pioneering scientist in atmospheric numerical modeling. He attended Dartmouth College and received his B.A. in general humanistic studies in 1937; then in graduate school, he earned an M.S. in geology from Columbia University in 1942 and a Ph.D. in meteorology in 1949 from UCLA, only the second to receive such a degree at that school. He stayed at UCLA to begin his career, working with Jacob BJERKNES, who had founded the UCLA meteorology department only eight years earlier. Mintz worked as his assistant on the UCLA General Circulation Project of the Atmosphere and later became the chair of the department of meteorology (1967–69). In the mid-1950s, the weather models used by forecasters were still regional or continental (vs. hemispherical or global) in scale, but Mintz and other theoretical meteorologists—unconcerned with real-time forecasting—found general circulation modeling a promising research tool.

Mintz recruited a Japanese mathematician, Akio Arakawa, to help him build general circulation models, and together they constructed a series of increasingly sophisticated models starting in 1961. The first UCLA general circulation model, now known as the Mintz-Arakawa Model, was a global one and was the first of its kind ever developed. The creating of models at UCLA had the most influence in the field, focusing on model development and letting other organizations conduct the experimental studies to test them. Their models, or evolutions of them, are still being used throughout the world for weather and climate predictions, as well as for scientific studies focusing on the global atmosphere.

In 1977, he retired from UCLA and joined the NASA/Goddard Space Flight Center, where as a consulting scientist he continued his research with efforts on understanding land-surface processes that fashion land-surface habitats and determine their biological productivity and response to human or natural environmental changes and the hydrological cycle. The hydrological cycle describes the circulation of water around the Earth, from ocean to atmosphere to the Earth's surface and back to the ocean. Mintz recruited young Englishman and biometeorologist Piers SELLERS to work for him in computer modeling of climate systems. Sellers went on to become an American astronaut.

Mintz was the author of many pioneering articles that became standard references on large-scale atmospheric and oceanic general-circulation models that led to significant understanding of climate processes. He coauthored several classic papers on the effect of the winds and tides on the variation of the length of the day and on the numerical simulation of stratospheric ozone. A major issue affecting the stratosphere is the depletion of the ozone layer, which leads to the formation of the "ozone hole" over the Antarctic in southern spring. His research has become standard reference material in textbooks.

He received the Clarence Leroy Meisinger Award from the American Meteorological Society in 1967, the NASA Award for distinguished service in 1988, and the Carl-Gustaf Rossby Research Medal from the American Meteorological Society in 1990. He died of cancer while in Israel on April 27, 1991.

⊠ **Mitseva-Nikolova, Rumyana Petrova**
(1948–   )
Bulgarian
Physicist

Rumyana Mitseva was born on October 30, 1948, in Dolna Banya, a small village in Bulgaria, to Peter Ivanov Mitzev, a teacher and researcher in geodesy and cartography, and Elissaveta Kostadinova Mitzeva, an accountant. As a youngster attending school in Sofia, Mitseva was interested in literature, theater, films, dance, and pantomime. While in school and as a student, she performed in nonprofessional theatre, took a part in a nonprofessional pantomime group, and competed in a dance club for competition (Latin and classic) dances. She attended the University of Sofia and received her B.S. in physics, specializing in solid state physics, in 1971 and, in 1975, an M.S. in meteorology. She married Peter Stefanov Nikolov in 1975, and they have one daughter. She received a Ph.D. in physics in 1987. At that point, she began her career as a physicist in the laboratory of semiconductors in the Institute of Physics in Sofia (As a rule in the eastern countries at the time, young people did not work after earning a degree if they continued to study in the university).

In 1978, she published her first scientific paper in Russian in the *Bulgarian Geophysical Journal*, entitled "Simulation of real powerful convective clouds and precipitation by a model of an isolated thermal" with her colleague Vassil Andreev; this paper checked the possibility of a one-dimensional numerical model to be used for the simulation of convective clouds, depending on the initial conditions and sounding.

Mitseva's main discoveries are in the field of cloud physics, namely numerical study of the formation of precipitation and thunderstorm electrification. She discovered relationships between the type of precipitation (rain, hail, or heavy hail) at a particular area in Bulgaria and in-cloud characteristics estimated by a numerical cloud model. These relations were used to forecast precipitation and hail intensity in Bulgaria. Some of her numerical studies show that the positive/negative dynamical effect due to the seeding of clouds for rain enhancement may lead to the decrease/increase of the rain.

In the field of thunderstorm electrification, her studies attempted to understand better the microphysical controlling variables that influence charge transfer and to build the laboratory

Rumyana Petrova Mitseva-Nikolova. Mitseva's main discoveries are in the field of cloud physics, namely numerical study of the formation of precipitation and thunderstorm electrification. *(Courtesy of Rumyana Petrova Mitseva-Nikolova)*

data into a model of thunderstorm electric field growth, in accordance with field observations.

Mitseva's numerical experiments are directed to the study of contribution of noninductive mechanism on the distribution of charge density in thunderstorms. Based on measurements of electric field changes associated with lightning, C. T. R. Wilson (1916) suggested a simple dipole structure for the charge distribution in thunderstorms, with positive charge above negative. Based on field measurements in different types of clouds, G. Simpson and Scrase (1937) and Simpson and C. T. R. Robinson (1941) suggested a

tripole charge structure that includes a lower region of positive charge. The numerical experiments showed that the so-called noninductive mechanism is able to explain the formation of two or three charge layers and that this mechanism plays the most significant role in thunderstorm electrification at least during the growth stage of clouds.

However, W. David Rust and Thomas C. Marshall (1996) recently questioned the analysis of the tripole model, basing their analysis on balloon-borne studies of electric fields inside thunderstorms by Marshall and Rust (1991), S. M. Hunter et al. (1992), Marshall and Rust (1993), and Marshall et al. (1995). Maribeth STOLZENBURG (1998) showed vertical charge profiles through thunderstorms with several layers of alternating charge sign; they conclude that the electrical structure of most thunderstorms is more complicated than the tripole model. These new measurements raise questions concerning the particle-charging mechanisms, which so far have been invoked to explain only the tripole charge structure. Stolzenburg et al. in 1998 suggested that several particle-charging mechanisms are needed to account for the observations of multiple charge layers alternating in polarity.

A more recent numerical study made by Mitseva and colleagues showed that a noninductive mechanism alone is able to account for the detection by balloon sounding of more than three charge layers alternating in polarity. Although the magnitude of charge is small when the interaction of ice crystals and graupel takes place in the absence of cloud droplets, it affects the charge distribution in the thunderstorm when the temperature at cloud-top height is at very low.

Mitseva's discoveries help to understand more details of the formation of precipitation in natural and seeded clouds and to contribute to a better understanding of thunderstorm electrification, which is still unresolved. She has published about 30 scientific articles and received a number of grants from the Royal Society and

Bulgarian Academy of Sciences which have permitted her to visit UMIST (University of Manchester Institute of Science and Technology) since 1989 to work with her colleagues in the field of thunderstorm electrification. She also received grants from the Bulgarian foundation (National Science Foundation and Science foundation of Sofia University) to work in the field of cloud physics.

Mitseva's most important contribution is in the field of thunderstorm electrification. Based on the laboratory work at the University of Manchester, Institute of Science and Technology (UMIST), done by Clive P. R. Saunders and his Ph.D. students, laboratory studies showed that the interaction of ice crystals with graupel pellets collecting supercooled water droplets can separate electric charge, leading to the development of high electric fields and lightning in thunderstorms. This mechanism is called noninductive

mechanism because the charge is transferred independently of the local electric field. It is believe that this mechanism plays the most important role (among other mechanisms) in thunderstorm electrifications.

The experiments established that the sign of the charge transferred between ice particles depends on the temperature and liquid water content in the cloud. Based on the data from laboratory experiments carried out by Saunders and his students, Mitzeva derived empirical equations that give opportunity to calculate the magnitude and sign of the charge transfer at rebounding collisions of ice crystals and graupel as a function of liquid water content and in-cloud temperature. The equations were used in numerical models to study electrification of thunderstorms. She continues her research as a associate professor in the department of meteorology, faculty of physics, at the University of Sofia.

N

Neumann, John Louis von (Johann)
(1903–1957)
Hungarian/American
Mathematician

Johann von Neumann (later changed to John) was born on December 28, 1903, in Budapest, Hungary. He was the oldest of three sons of Max Neumann, a banker, and Margaret Kann Neumann. Having a Jewish background, they had to flee Hungry to Austria when communists took over briefly in 1919. At an early age, he was privately tutored, and evidence of his genius was exhibited by the age of six when he could divide eight-digit numbers in his head.

From 1911 to 1916, Neumann attended the Lutheran gymnasium in Budapest, in 1921, enrolled at the University of Budapest, and also attended the University of Berlin from 1921 to 1923, studying chemistry and listening to lectures by Albert Einstein. In 1923, he published a definition of ordinal numbers, still in use today. He received a degree in chemical engineering at the Swiss Federal Institute of Technology in Zurich in 1925 and, the following year, his doctorate in mathematics, with minors in chemistry and experimental physics, from the University of Budapest. His thesis was on set theory.

While studying as a Rockefeller Fellow at the University of Göttingen, in 1927, he met J. Robert Oppenheimer, who would become a leading figure in U.S. nuclear research at Los Alamos, New Mexico, during World War II. Neumann accepted a lecturer position in mathematics at the University of Berlin from 1927 to 1930. While in Berlin, he published a number of papers that would have major impact in the field of quantum theory and economics.

Neumann married Marietta Kovesi in January 1930, and the couple had one daughter (but divorced in 1937). After his marriage that year, he came to America and became a visiting lecturer at Princeton, but within a year, he was a tenured professor, lecturing on mathematical hydrodynamics. Two years later he became one of the six original professors at the university's newly formed Institute for Advanced Study.

In 1932, he published *Mathematische Grundlagen der Quantenmechanik* that laid the mathematical foundation of quantum mechanics that was founded earlier in 1925. From 1933 to 1935, he took on coediting responsibilities for the *Annals of Mathematics* and *Compositio Mathematica*, holding both positions until his death.

Neumann became a U.S. citizen in 1937, in time to receive wartime security clearance and to begin to work for the U.S. government on the

John von Neumann. His work on parallel processes and networks has earned him the label of the "father of the modern computer." *(Alan Richards, photographer. Courtesy of the Archives of the Institute for Advanced Study)*

Manhattan (atomic bomb) Project at Los Alamos, particularly on detonation devices, along with J. Robert Oppenheimer. In 1938, the American Mathematical Society awarded him the Maxime Bôcher Prize for his memoir "Almost periodic functions and groups." At the end of the year, in December, he married again, to Klara Dan; they lived in Princeton and then later in Georgetown, Washington, D.C.

After World War II, Neumann turned his attention to improving the speed of computers and their ability to perform computations. He is credited as the first person to endorse publicly the creation of a computer that had memory capable of storing programs. In 1944, Neumann and friend Oskar Morganstern published a book titled

Theory of Games and Economic Behavior. They analyzed such strategy games as poker, showing that a "best possible" method of play existed and could be determined using math. This became known as game theory and provided ideas and techniques that could also be applied to economic, social, and military problems.

By 1944, he was corresponding with Howard Aiken (Harvard Mark I), George Stibitz at Bell Labs, and others, and by a chance meeting with Herman Goldstine in June, he learned about the work being done at the Moore School of Electrical Engineering at the University of Pennsylvania. Researchers there were building a computer called the Electronic Numerical Integrator and Computer (ENIAC) that could perform more than 300 multiplications per second. Neumann received clearance in September to visit the school and quickly became a consultant to the project. He immediately began to research for an improved computer called the Electronic Discrete Variable Automatic Computer (EDVAC).

Neumann wrote a 101-page report called the First Draft of a Report on the EDVAC in 1945 that described the theory and design of the EDVAC, informing readers that it would have a memory and store data and programs. Unfortunately, outsider researchers erroneously believed that the whole concept was Neumann's and not his other colleagues, one of whom had written a memo detailing the concept six months before Neumann even heard of ENIAC. Goldstein and Neumann applied for a patent from the Pentagon on March 22, 1946, but was refused on the basis that patents had to be filed within a year after the appearance of published data.

However, it was Maurice Wilkes, at Cambridge University Mathematical Laboratory, who conceived his own design for the EDSAC (Electronic Delay Storage Automatic Calculator); that became the world's first operational stored-program computer. EDSAC performed its first calculation on May 6, 1949.

Neumann was awarded the Medal for Merit and the Distinguished Civilian Service Award in 1946 for his outstanding contributions during World War II and the Medal of Merit Presidential Award in 1947. In 1948, Neumann published *Cybernetics: Or Control and Communication in the Animal and the Machine*, promoting the idea that electronic "brains" could perform human tasks.

Neumann stayed at Princeton and continued to work on computer theory, and his research of stored-program computers, which became universally adopted, is now an integral part of modern high-speed digital computers. He convinced Princeton to allow him to build his own computer, the IAS (named for his place of research, the Institute for Advance Study). It was completed in 1951. The following year, he designed MANIAC 1 (mathematical analyzer, numerical integrator, and computer), a high-speed computer designed to use a flexible stored program. Historians agree that his MANIAC was used to produce and test the world's first hydrogen bomb in 1952.

From 1948 to 1956, Neumann was supervisor to Jule Gregory CHARNEY, who was also a member of the Institute for Advanced Study in Princeton. Neumann believed that the goal of weather prediction should have the highest priority in the use of his computer. Charney served as the director of theoretical meteorology for the AIS group. Using ENIAC and MANIAC, he and the group constructed a successful mathematical model of the atmosphere, demonstrating that numerical weather prediction was both feasible, although it took 24 hours to generate the forecast on ENIAC and five minutes on MANIAC. This satisfied the theories laid down earlier by BJERKNES, L. F. RICHARDSON and C. G. ROSSBY. Charney went on to help establish a numerical weather prediction unit within the U.S. Weather Bureau in 1954.

In October 1954, President Dwight D. Eisenhower appointed him a member of the Atomic Energy Commission (AEC). The AEC, established after World War II, was responsible for developing and producing nuclear weapons and for regulating the use of atomic power in industry in the United States, although Neumann was politically an independent. Neumann's work also led to the construction of ORDVAC (ordnance variable automatic computer), unveiled in 1952. He was appointed a member of the General Advisory Committee of the AEC. In 1954, a new calculator for the navy, NORC (naval ordnance research computer), was boasted to do a 24-hour weather prediction in a few minutes. The New York *Herald Tribune* (December 3, 1954) also reported that NORC could compute the tidal motions of the entire Atlantic and Pacific Oceans; throw new light on the core of the Earth, believed to be liquid, by computing the turbulent motion at the center; and help the armed forces plan the movement of men and material by mathematically simulating logistical problems.

His later work on parallel processes and networks has earned him the label of the "father of the modern computer." During the summer of 1954, he learned that he had bone cancer that may have been the result of exposure to nuclear radiation during his work at Los Alamos.

In 1956, he received the Albert Einstein Commemorative Award, the Enrico Fermi Award, and the Medal for Freedom. Also that year, Neumann, Charney, and others served on a National Science Foundation committee to develop a national meteorological policy that would promote research, education, and professionalism. In 1960, the National Center for Atmospheric Research, NCAR, was established. Even near death, he met with scientists and government officials on the issue of the AEC at his hospital bedside. On February 8, 1957, Neumann died from cancer at the age of 53. He is considered one of the greatest minds of the 20th century.

Besides his awards and medals, he received honorary doctoral degrees from Princeton (1947), Harvard University (1949), Istanbul University, Turkey (1952), University of Munich, Germany (1953), and Columbia University

(1954). He was a member of many professional societies: American Mathematical Society (president 1950–51), the National Academy of Sciences (1937–57), American Philosophical Society, Mathematical Association of America, American Academy of Arts and Sciences, Academiz Nacional de Ciencias Exactas, Lima, Peru; Acamedia Nazionale dei Lincei, Rome, Italy; Instituto Lombardo di Scienze e Lettere (Milan, Italy); Royal Netherlands Academy of Sciences and Letters, Amsterdam, Netherlands; and Information Processing Hall of Fame, Infomart, Dallas TX, 1985.

⊠ Newton, Sir Isaac
(1643–1727)
English
Mathematician, Physicist

Isaac Newton was born on January 4, 1643, in Woolsthorpe, Lincolnshire, to Isaac Newton, a farmer who died three months before the famed mathematician was born, and Hannah Ayscough. She later remarried to Barnabas Smith, the minister of a church at nearby North Witham, when Newton was two years old, and his grandmother Margery Ayscough at Woolsthorpe raised him. Although Newton was destined to become one of the greatest scientists of all time, he was considered a controversial figure in his day due to his relationships with his peers.

Newton first lived with his mother and an extended family after his stepfather died in 1653. He attended the Free Grammar School in Grantham but showed little enthusiasm for academics and was taken out to run his mother's estate. That also proved of no interest to him. He returned to the Grammar School in 1660, on the convincing of an uncle, to finish and attend a university. Newton entered Trinity College Cambridge in 1661.

Newton studied ARISTOTLE, Descartes, Gassendi, Hobbes, Boyle, GALILEO, and Kepler but it

Sir Isaac Newton. Controversial but brilliant, Newton is given credit for founding the laws of gravity, inventing a new type of telescope, and spending his later years in politics. *(Courtesy AIP Emilio Segrè Visual Archives)*

was clear that he was forming his own ideas at an early age. He also began to read many of the math books of the time. He received his bachelor's degree in 1665 and returned to Lincolnshire when a plague closed down the university that summer. Returning later for a master's degree, Newton was elected a Fellow of Trinity College in 1667, and when his teacher Isaac Barrow resigned from the Lucasian Professor chair in 1669, Newton took it over on Barrow's recommendation.

Newton's first task was to work in the field of optics. It was widely held that white light was a single entity, but Newton had decided that it was not, based on chromatic aberration he found in a telescope lens. Newton conducted an experiment of passing a thin beam of sunlight through a glass prism and observing the resulting spectrum of colors. He argued that white light was made up of different types of tiny corpuscles that refracted at different angles and that each produced a different color. Opposed to the construction of refracting telescopes because he erroneously thought that all refracting lenses would have chromatic aberration, he instead designed a reflecting telescope in 1672, to this day called a Newtonian. He also invented the sextant, which was rediscovered by John Hadley and Thomas Godfrey independently of each other in 1731.

In 1672, Newton was elected a Fellow of the Royal Society and published his first scientific paper on light and color in their *Philosophical Transactions of the Royal Society*. Robert HOOKE and Christiaan Huygens, however, disagreed that light was the motion of small particles: this began the lifelong feud between Newton and Hooke. To compound the feud, in 1675, Hooke argued that Newton had stolen some of his conclusions regarding light, so Newton waited until Hooke died in 1703 before he published his *Opticks* (1704). It was not until the 19th century that the wave theory for light, an explanation of optical phenomena, was revived by scientists such as François Arago, Augustin Fresnel, James Clerk MAXWELL, and Heinrich HERTZ. In 1678, Newton suffered the first of two nervous breakdowns as a result of arguments with English Jesuits in Liège who disagreed with his theory of color.

Newton's greatest achievement is his theory of universal gravitation: "All matter attracts all other matter with a force proportional to the product of their masses and inversely proportional to the square of the distance between them." In 1687, Newton published *Principia*, or *Philosophiae naturalis principia mathematica (Mathematical Principles of Natural Philosophy)*, which some historians proclaim is the greatest scientific book ever published. His physics not only applied to celestial bodies but was also relevant to the trajectory of projectiles, pendulums, and free fall such as the proverbial apple falling from the tree.

In *Principia*, he demonstrated that the planets are attracted toward the Sun and that all celestial bodies mutually attract each other. Newton went on to explain the eccentric orbits of comets, tides and their variations, the precession of the Earth's axis (the Earth's axis rotates [precesses] just like a spinning top. The period of precession is about 26,000 years), and lunar motion as it is affected by the Sun's gravity. Volume I of *Principia* lays the framework for the science of mechanics, while volume II puts forth the theory of fluids, and volume III describes the law of gravity.

After the crowning of William and Mary in 1689, Newton was elected to Parliament for his staunch stand against making universities Catholic (he was Protestant). Shortly after, in 1693, Newton suffered his second nervous breakdown, left Cambridge (finally resigning in 1701), and took a government job in London, becoming warden of the Royal Mint in 1696 and master in 1699. Now quite wealthy, Newton concentrated on the issues of the mint, dealing with coinage and preventing counterfeiting. He was again elected to Parliament in 1701–02.

In 1703, he was elected president of the Royal Society and was reelected each year until his death. In 1705, he was the first scientist ever knighted and is considered one of the three most important scientists of all time.

Newton ended his days once again in controversy, this time with Gottfried Wilhelm von Leibniz over who invented calculus. As president of the Royal Society, Newton put together a committee to arbitrate, but he also wrote anonymously the official report supporting himself, along with an anonymous review of it in the *Philosophical Transactions of the Royal Society*. He remained master of the mint until his death on March 31, 1727, in London, where he was buried with much pomp and ceremony in Westminster Abbey.

O'Brien, James Joseph Kevin
(1935–   )
American
Meteorologist, Physical Oceanographer

James O'Brien was born on August 10, 1935, in the Bronx, New York, to Maurice O'Brien and Beatrice Cuddihy, who met on a boat from Ireland in 1927. Although Maurice had no education beyond sixth grade, he was made a metallurgist technician at Bell Labs because of his cleverness and eventually moved to their Murray Hill, New Jersey, labs. As the oldest child in a family of nine children, James O'Brien loved math and loved to read. Though he lived in New Providence, he attended Summit High School in Summit, New Jersey, where 85 percent of his class went to college. O'Brien went to Rutgers Men's College in 1953 and graduated with a bachelor's degree in chemistry in 1957. He also received a commission in the U.S. Air Force as a second lieutenant. He married Sheila O'Keefe, and they had five children.

O'Brien first worked for E. I. DuPont de Nemours and Company, Inc., as a chemist in 1957, until the air force sent him to the University of Texas at Austin for 12 months to learn meteorology. He was then stationed at Bolling Air Force Base in Washington, D.C., for 15 months as a weather officer but was discharged early. He went back to DuPont. In 1962, he moved with his family to Texas A&M to get his Ph.D. He completed both a master's degree (1964) and Ph.D. (1966) in meteorology in three years and one semester, a record that still stands at the university.

While in graduate school at Texas A&M, he took a statistics course with John Griffiths. As a result of asking some questions, O'Brien wrote two papers, "The rank correlation coefficient as an indicator of the product-moment correlation coefficient for small samples (10–100)," published in the *Journal of Geophysical Research* (1965), and "On choosing a test for normality for small samples," published in *Archiv für Meteorologic, Geophysik und Bioklimatologie* (with John F. Griffiths in 1967). His master's degree thesis, entitled "Investigation of the diabatic wind profile in the atmospheric boundary layer," was published in the *Journal of Geophysical Research* in 1965. This paper explained the structure of near-surface wind profiles based on a Russian theory.

O'Brien has held several positions, including Advanced Study Group Fellow and research scientist at the National Center for Atmospheric Research, in Boulder, Colorado (1966–68). From 1969 to the present, he has held positions at Florida State University as first associate professor and then professor of meteorology and

James Joseph Kevin O'Brien. In the field of meteorology, he was the first to study the effect of a hurricane on the ocean, and this started an entire field of study. *(Courtesy of James Joseph Kevin O'Brien)*

oceanography. He currently is the Robert O. Lawton Distinguished Professor of meteorology and oceanography and director of the Center for Ocean–Atmospheric Prediction Studies.

In the field of meteorology, he was the first to study the effect of a hurricane on the ocean, and this started an entire field of study. He discovered the way to calculate vertical velocity from typical atmospheric wind profiles soundings and has more than 4,000 citations proving that this is the way to do it. He has determined how to calculate the vertical profile of horizontal eddy diffusivity and now has more than 2,000 citations. This idea is even used in the National

Center for Atmospheric Research ocean model to handle convection in the ocean.

He has created a number of models of El Niño, among them a model that demonstrated Klaus WYRTKI's ideas of how Kelvin Waves from the western Pacific create El Niño in the eastern equatorial Pacific; a real model of El Niño showing how real winds give the amplitude and phase of all Los Niño's; and effective forecast models of El Niño. Other discoveries include reology, or the stress-strain relationship of marginal ice zones that shows why the marginal ice zone is thin and how ice-edge upwelling occurs, and ENSO (El Niño—Southern Oscillation) impacts over the United States (tornadoes, hurricanes, snow, floods, and high winds) through a series of papers written jointly with his students finding ENSO signals in the historical climate records.

O'Brien is a Fellow of the American Association for the Advancement of Science (1988), American Geophysical Union (1987), American Meteorological Society (1981), and Royal Meteorological Society (1983); a Foreign Fellow, of the Russian Academy of Natural Science (1994); and a member of the Oceanographical Society of Japan, The Norwegian Academy of Science and Letters, and other societies. His numerous awards and citations include the Medal of Honor of Liège University, Belgium (1978); Sverdrup Gold Medal in Air–Sea Interaction (1987); ONR Distinguished Ocean Educator (1989); and Medal of Honor, Ocean University of Quindao, China (1999). He is the author of more than 200 scientific papers and reports and has served as associate editor for *Monthly Weather Review* (1992–97); *Continental Shelf Research* (1986–   ); and for *International Journal of Math and Computer Modeling* (1984–   ). He has been a member of numerous committees and panels and reviewer for a number of scientific organizations.

O'Brien has trained 34 Ph.D. students, 45 master's degree students in meteorology and physical oceanography, and more than 25 postdoctoral students. He and his students have convinced the field that isopycnal vertical coordinates or constant density surfaces are the only way to understand the ocean.

O'Brien's major contributions, however, are in solving the vertical velocity problem and the ENSO problem, which is the vertical motion of the air or water that is important for rain or life.

It is very small and hard to measure directly. His formulae give good estimates. The understanding of the fluid dynamics of ENSO events has led to an ability to forecast El Niño and La Niña several months in advance. He is continuing his work as professor of meteorology and oceanography, at the Center for Ocean-Atmospheric Prediction Studies at Florida State University.

P

Philander, S. George H.
(1942–)
South African
Oceanographer

George Philander was born in South Africa to Peter and Alice Philander, both high school teachers. Philander attended a segregated school during this time of apartheid; however, when he attended the University of Cape Town, it was not segregated and he received his B.A. there in applied mathematics in 1962. He then moved on to Harvard, where he obtained his Ph.D. in applied mathematics in 1970. He married Hilda Storari, a meteorologist from Argentina whom he met in graduate school; they have one son.

After accepting a postdoctoral fellow position at Massachusetts Institute of Technology (MIT) in 1970–71, Philander became a research associate in the geophysical fluid dynamics program at Princeton University, New Jersey, from 1971 to 1977, serving as a consultant to the World Meteorological Organization in Geneva, Switzerland, for a year (1973–74). From 1978 to 1989, he was a senior research oceanographer in the Geophysical Fluid Dynamics Laboratory at the National Oceanic and Atmospheric Administration (Princeton, New Jersey). By 1990, Philander was back at Princeton University as a professor of geological and geophysical sciences

and from 1994 to 2001 was chairman of the department.

Philander's contributions have been mostly in oceanography and in the field of ocean–atmosphere interactions, which concern phenomena such as the Southern Oscillation between complementary El Niño and La Niña states. (He introduced the latter term). He contributed significantly to the development of computer models that simulate the oceanic circulation, the counterparts of the models used to predict weather.

He has served on a number of committees and panels and has authored more than 100 scientific publications including papers, chapters for several books, and two books dealing with the issue of global warming: *El Niño, La Niña and the Southern Oscillation,* (1990) and *Is the Temperature Rising? The Uncertain Science of Global Warming* (1998). The first is intended for researchers in the field of ocean–atmosphere interactions. The second, intended for lay readers, explains some of the physical and chemical processes that make ours a habitable planet and explores the likely consequences of the current rapid rise in the atmospheric concentration of greenhouse gases.

Philander has received a number of awards including NOAA Environmental Research Laboratories, Distinguished Authorship Award (1979 and 1983), the Sverdrup Gold Medal of the

America Meteorological Society (1985) to which he was elected a Fellow, and the Department of Commerce Gold Medal (1985). In 1991, he was elected a Fellow of the American Geophysical Union, and three years later, he was invited to give the Symons Memorial Lecture of the United Kingdom's Royal Meteorological Society. He is currently conducting research in the area of paleoclimates—phenomena such as the recurrent ice ages. He reasons that confidence in climate models that predict future global warming will be bolstered considerably if those models can explain the very different climates the Earth has experienced during its long and fascinating history.

R

⊠ **Rakov, Vladimir A.**
(1955–   )
American
Electrical engineer

Vladimir Rakov was born on August 7, 1955, in Semey (Semipalatinsk), Kazakhstan, to Aleksandr I. Rakov, an electrical engineer, and Lidiya D. Yakovleva, a physician. As a youngster he enjoyed reading books and stamp collecting while he went to high school in Semey, graduating with honors in 1972. In 1974, he married Lioudmila D. Fateeva, with whom he has a son. Rakov went on to Tomsk Polytechnic University in the former USSR and received a master's degree in electrical engineering in 1977 with his thesis "Multi-criteria optimization of the parameters of the Nurek-Regar 500-kV power transmission line." This study was designed for finding the design parameters of the power line that best satisfy a set of criteria, some of which are in conflict with each other (for example, the minimum investment and the maximum reliability). He received a Ph.D. in electrical engineering in 1983 on the theisis "Development of techniques for the determination of lightning peak current statistical distribution."

Rakov first published on the multicriteria optimization of power lines in 1978 while assistant professor of electrical engineering at Tomsk Polytechnic Institute. However, his research interests moved to the lightning area. He developed a theory of the behavior of lightning once it has hit the ground and determined the basic modes of lightning-charge transfer to the ground. His theory improved understanding of the physics of lightning discharge and its interaction with the environment. He has obtained more than $3.5 million in external funding for his research.

In 1991, after directing for seven years the Lightning Research Laboratory at Tomsk Polytechnic, Rakov became associate professor in the department of electrical engineering, University of Florida at Gainesville. He has published more than 260 papers, and his major contributions deal with the understanding of the physics and effects of lightning. He is the lead author of the 850-page monograph *Lightning: Physics and Effects* by Rakov and Uman, and he has more than 30 inventions dealing with lightning.

Throughout his career, his peers have recognized him. In 2001, he was awarded the Institute of Electrical and Electronic Engineers (IEEE) Power Engineering Society Surge Protective Devices Committee, Prize Paper Award. He received the University of Florida Research Foundation Professorship Award for 2001–03. He is also listed in *Who's Who in America* and *Who's Who in Science and Engineering*. In 1999, the American Society for Engineering Education

Vladimir A. Rakov. His major contributions deal with the understanding of the physics and effects of lightning, and he has more than 30 inventions dealing with lightning. *(Courtesy of Vladimir A. Rakov)*

(ASEE) awarded Rakov the Southeastern Section Medallion Certificate, Research Unit Award for Outstanding Contribution in Research. The Tomsk Polytechnic Institute gave him the rank of senior scientist in 1985 and named him the best young scientist and best young inventor within educational and research institutions in Tomsk in 1981, 1985, and 1988. In 1986, USSR State Committee for Inventions and Discoveries gave him the medal "Inventor of the USSR," and the following year, the Main Committee of the USSR Exhibition of Technological Achievements awarded Rakov the Silver Medal.

Rakov is a member of the American Geophysical Union (1989) and the American Meteorological Society (1996) and is a senior member of IEEE (1996), the Society of Automotive Engineers (SAE Aerospace, 1999), the IEEE Power Engineering Society (2001), and the EMC Society (2001). He has served as chair or member on a number of committees and given numerous lectures on lightning issues and research.

In 1998, he became full professor in the department of electrical and computer engineering at the University of Florida in Gainesville, a post he still holds today. Rakov has mentored more than 50 students working on their graduate theses. He continues to work in the field of lightning research.

⊠ Rasmusson, Eugene Martin
(1929–   )
American
Meteorologist

Eugene Martin Rasmusson was born on a farm in McPherson County, Kansas, on February 27, 1929, the first of seven children of Martin Erick and Alma Sophia (Nelson) Rasmusson. His father was a farmer, with wheat as the major cash crop. Farm life in central Kansas during Rasmusson's childhood was quite austere, with no electricity, no indoor plumbing, and a telephone only when it could be afforded. His first decade of life also coincided with the wrenching drought and dust-bowl conditions of the 1930s. He retains a vivid memory of the consequences of this climate catastrophe on the welfare of the farm families of his community. The crucial nature of weather and climate for the local farmers was a major factor in stimulating his interest in meteorology and ultimately in determining his future career.

Rasmusson received his primary education in a one-room rural school. One teacher, with rarely more than two years of college training, taught all grades in which there were students. This rarely included all eight grades because there were never more than a dozen students in the school. The school year was eight months long, and like most of the other farm children in the area, Rasmusson spent the summers helping with the farmwork. His horizons were greatly expanded when at age nine the family acquired their first 6-volt battery powered radio. This was his window to the world and introduced him to the fascinating world of science. One of his early memories is of a radio program that dealt with science and history topics, sponsored by DuPont. He particularly remembers their slogan "Better things for better living through chemistry" and an episode that attempted to explain relativity. He remembers thinking what an exciting life it would be to be able to leave the farm and become a scientist.

Rasmusson attended high school (1942–46) in the neighboring town of Lindsborg, Kansas (population around 2,000), and on graduation enrolled in the civil engineering curriculum at Kansas State University in Manhattan, Kansas. During his four years at Kansas State, he also completed the advanced course in air force ROTC, receiving a reserve commission as a second lieutenant on graduation in 1950. There were no formal courses in meteorology at Kansas State at that time, but his interest in the atmospheric sciences was greatly stimulated by an introduction to meteorology and weather forecasting taught as part of his ROTC training. His graduation coincided with the beginning of the Korean War, and after working for only nine months as a surveyor with the Kansas Highway Commission, he was called to active duty in 1951. He accepted an opportunity to pursue a one-year basic meteorology course at the University of Washington (1951–52) as preparation for his assignment as an air force weather forecaster. This decision was to influence the course of his professional life profoundly. Following this training, and three years of service as an air force forecaster at bases in Oklahoma and Alaska, he was released from active duty in 1955. He remained a member of the Air Force Active Reserve, retiring as a lieutenant colonel in 1974.

Returning to civilian work, Rasmusson accepted a position as a plant engineer with Pacific Telephone and Telegraph Co. in Seattle, Washington. He quickly realized however that meteorology and not civil engineering had become his consuming professional interest, so he resigned this position in 1956 to accept a position as a U.S. Weather Bureau river forecaster in St. Louis. His subsequent four years of work in hydrology and river forecasting turned out to be a valuable asset in his future career in climate research. It was while working in this position that he met Georgene Sachtleben, a music teacher, and the two were married in August 1960. They have four children. In 1960, he published his first scientific paper, "Extended Low Flow Forecasting Operations on the Mississippi River" in the *I. A. S. H. Commission on Surface Waters Publication*, and was promoted to a state weather forecaster for eastern Missouri and Southern Illinois.

During his eight years in St. Louis, Rasmusson took graduate-level night courses at St. Louis University. Because only engineering and applied mathematics courses were offered in the night-school program, he pursued a major in engineering mechanics, receiving a master's degree in 1963. A few months later, he was awarded a U.S. Weather Bureau scholarship to pursue Ph.D. studies at MIT. Accepting this opportunity, he studied under one of the premier meteorological faculties of the time and interacted with a highly talented group of graduate students. The three years at MIT marked a shift in his professional career from operational forecasting to research.

Following receipt of his Ph.D. in meteorology in 1966, Rasmusson joined the staff of the Geophysical Fluid Dynamics Laboratory (GFDL) in Washington, D.C. GFDL was the premier general-circulation and climate-modeling laboratory of the time, and he and his colleague Dr. A. H. Oort set as their major research project the comprehensive documentation of the general circulation of the atmosphere, using the first multiyear computer database of atmospheric upper-air observations. Their research culminated in the publication of a comprehensive monograph, *Gen-*

Eugene Martin Rasmusson. Rasmusson's perceptive analysis of the El Niño/Southern Oscillation (ENSO) phenomenon, which is the largest single source of interannual global climate variability, brought him worldwide recognition and provided information needed for the development of coupled-ocean/ atmosphere-prediction models. *(Courtesy of Eugene Martin Rasmusson)*

eral Circulation Statistics, which served as the internationally known "bible of the general circulation" for more than a decade. It was the primary information source for a multitude of empirical studies, as well as the "ground truth" against which to compare the results of newly developed global climate models.

Rasmusson has consistently approached the problems of climate from a coupled land/ocean/ atmosphere perspective. This broad, interdisciplinary approach was already apparent in his Ph.D. research on the hydrologic cycle of North America, which broke new ground by analyzing the surface and atmospheric branches of continental-scale hydrology not as independent components but rather as interacting elements of a coupled system. This study was in many ways ahead of its time and has been recognized during the past two decades as a seminal contribution to the analysis of continental-scale hydrology. This approach to hydrologic analysis is now a central element of the water-balance studies being conducted in several regions of the world as part of the Global Energy and Water Cycle Experiment (GEWEX).

Rasmusson accompanied GFDL in their move to Princeton University in 1968 but in 1970 returned to Washington. He was asked to lead a newly formed National Oceanic and Atmospheric Administration (NOAA) group, tasked with interpreting the data collected during BOMEX, the first large-scale ocean–atmosphere experiment, conducted in 1969 near the island of Barbados. In recognition of his definitive determination of the fluxes of moisture, energy, and momentum from ocean to atmosphere over the 500-km-square experimental array, he was awarded the U.S. Department of Commerce Silver Medal (1973).

From 1973 to 1979, Rasmusson assumed crucial leadership roles in two major international programs. He played a major role in the design of the Global Atmospheric Research Program (GARP) Atlantic Tropical Experiment (GATE), which was undertaken over the eastern tropical Atlantic to provide data for improving the modeling of convection. As leader of the GATE Convection Subprogram Data Center, he was responsible for the assembly of observational data sets for this subprogram and participated in their analysis. During this period, he also played a major leadership role in the design and research activities of the Lake Meteorology Subprogram of the U.S.–Canadian International Field Year on the Great Lakes.

In 1979, Rasmusson organized the diagnostic branch of the newly formed NOAA Climate Analysis Center and refocused his personal research efforts on the role of the equatorial Pacific in climate variability. As chief of the diagnostics branch of the center, he initiated a monthly climate diagnostics bulletin that was the first internationally circulated summary and evaluation of the impacts of current climate anomalies throughout the world. In 1983 he received the NOAA Administrator's Award for his analysis of the catastrophic 1982–83 El Niño/Southern Oscillation (ENSO) episode and for his service to the public and scientific community in disseminating current information on the evolving anomalies and their impacts during this major climate catastrophe.

Rasmusson's perceptive analysis of the El Niño/Southern Oscillation (ENSO) phenomenon, which is the largest single source of interannual global climate variability, brought him worldwide recognition and provided information needed for the development of coupled ocean/atmosphere prediction models. This research was also crucial in laying the groundwork for the Tropical Ocean Global Atmosphere Experiment. This was a 10-year (1985–94) program element of the World Climate Research Program to further observe, study, and if possible predict the ENSO phenomenon. Rasmusson was one of the intellectual leaders in the conception and design of this program. In 1989, he received the Jule Charney Award of the American Meteorological Society for his major contributions to an understanding of this system of climate anomalies.

Rasmusson retired from NOAA in 1986 to become a research scientist at the University of Maryland, where he continued his research and leadership roles in international programs on climate variability and global/regional hydrology. He was instrumental in the development and research activities of the GEWEX Continental International Program (GCIP), an experiment designed to study the large-scale water and energy balances of the Mississippi Basin. He subsequently demonstrated that the North American warm-season precipitation regime could be viewed productively in a monsoon-system context. This led to the development of a new international climate research program focused on

climate and hydrologic variability over North and South America from a monsoon perspective.

Rasmusson has published seven book chapters and more than 100 scientific papers in a variety of refereed journals. These include invited contributions to the journals *Nature, Science, American Scientist,* and *Climate Change.* His book chapters include a chapter on climatology in the *Handbook of Hydrology* and a summary of the lectures he presented on the climate of the 20th century at the Les Houches 1977 summer school, *Modeling the Earth's Climate and its Variability.* He has contributed articles to *The Encyclopedia of Science and Technology* and its *Yearbook* and to *Science Year of the World Book.* His publications include papers and/or book chapters published by The National Academy Press, the World Meteorological Organization, the Pontifical Academy of Sciences, and the International Rice Institute. As previously noted, he and his colleague Dr. A. H. Oort published the first comprehensive monograph on the atmospheric general circulation. He has published more than 40 papers in conference and symposia proceedings and has written several less technical articles directed toward the nonspecialist and the secondary school audience. He has given scores of lectures to a variety of scientific, nonscientific, and student audiences.

Rasmusson has served the atmospheric sciences profession and the nation through membership on numerous advisory committees for the government and the National Research Council. He was chair of NASA Scientific Steering Group for the joint U.S.–Japanese Tropical Rainfall Measurement Mission (TRMM) satellite, which was launched in 1997. He has served on seven NRC committees and panels, including the Board of Atmospheric Sciences and Climate and chair of the Climate Research Committee. In addition, he has served on six international advisory committees dealing with climate and hydrological experiments.

Rasmusson has also assumed a leadership role in the American Meteorological Society. He has been an associate editor and editor of two AMS journals. He was president of the society in 1998. He has also served on several AMS committees, as a member of the AMS Council, and as chair of the AMS Planning Commission. In 2002, he received the C.F. Brooks Award for service to the society.

Rasmusson is a Fellow of the American Meteorological Society and the American Geophysical Union. His biographical sketches appear in *American Men and Women of Science, Who's Who in the World, Who's Who in Technology,* and *Men of Accomplishment,* among other lexica. Recognition for his career of research contributions and scientific leadership activities was capped in 1999 with his election to the National Academy of Engineering. In addition to the previously noted awards and recognition, Rasmusson guest lectured at the Iowa Academy of Science Annual Meeting (1984) and was the first Global Change Lecturer at the University of Arizona (1992). In 1994, he presented the annual Victor Starr Memorial Lecture at MIT, in 1998, he presented the George Benton Honorary Lecture at Johns Hopkins University, and was the American Meteorological Society Horton Lecturer in Hydrology in 1997. In 2000, Rasmusson became a research professor emeritus at the University of Maryland. Although formally retired, he continues to present lectures to a variety of audiences and remains active in NRC and AMS activities.

⊠ Réaumur, René-Antoine Ferchault de
(1683–1757)
French
Philosopher, Naturalist

René de Réaumur was born in La Rochelle, France, in 1683. After studying mathematics in Bourges, he moved to Paris in 1703 at age 20 and under the eye of a relative. Like most scientists of the time, he made contributions in a number of areas including meteorology. His work in mathematics allowed him entrance to the Academy of Sciences in 1708. Two years later, Réaumur was

put in charge of compiling a description of the industrial and natural resources in France and, as a result, developed a broad base view of the sciences. It also inspired him to invention that led him into the annals of weather and climate with the invention of a thermometer and temperature scale.

In 1713, Réaumur made spun glass fibers that today are the building blocks of Ethernet networking and fiber optics and are made of the same material. A few years later, in 1719, after observing wasps building nests, he suggested that paper could be made from wood in response to a critical shortage of paper-making materials (rags) at the time. He also was impressed by the geometrical perfection of the beehive's hexagonal cells and proposed that they be used as a unit of measurement.

In 1720, Réaumur developed the first malleable iron, known today as European Whiteheart. In his invention, the cupola furnace, he was perhaps the first to study the microstructure of iron and steel, showing their grainy structure. His cupola furnace would become a standard in iron foundries in the 19th century. He also improved techniques for making iron and steel, including tinning iron, and studied gold-bearing rivers and turquoise mines (showing that some turquoise was actually fossil teeth).

He turned his interests from steel to temperature and, in 1730, presented to the Paris Academy his study, "A Guide for the Production of Thermometers with Comparable Scales." Réaumur wanted to improve the reliability of thermometers based on the work of Guillaume AMONTONS, though he appears not to be familiar with FAHRENHEIT's earlier work.

His thermometer of 1731 used a mixture of alcohol (wine) and water instead of mercury, perhaps creating the first alcohol thermometer, and it was calibrated with a scale he created called the Réaumur scale. This scale had 0° for freezing and 80° for boiling points of water. The scale is no longer used today; however, most of Europe, with the exception of the British Isles and Scandinavia, adopted his thermometer and scale.

Unfortunately, his errors in the way he fixed his points were criticized by many in the scientific community at the time, and even with modifications in the scale, instrument makers favored making mercury-based thermometers. Réaumur's scale however, lasted for more than a century and, in some places, well into the late 20th century.

From 1734 to 1742, Réaumur wrote six volumes of *Mémoires pour servir à l'histoire des insectes* (Memoirs serving as a natural history of insects). Although unfinished, this work was an important contribution to entomology. He also noticed that crayfish have the ability to regenerate lost limbs and demonstrated that corals were animals, not plants. In 1735, he introduced the concept of growing degree-days, later known as Réaumur's Thermal Constant of Phenology. This idea led to the heat unit system used today to study plant-temperature relationships.

In 1737, Réaumur became an honorary member of the Russian Academy of Sciences, and the following year, he became a Fellow of the Royal Society.

After studying the chemical composition of Chinese porcelain, in 1740, he formulated his own Réaumur porcelain. In 1750, while investigating the animal world, he designed an egg incubator. Two years later, in 1752, he discovered that digestion is a chemical process by isolating gastric juice and studied its role in food digestion by studying hawks and dogs.

He died in La Bermondière, on October 18, 1757, and bequeathed to the Academy of Science his cabinet of natural history with his collections of minerals and plants.

⊠ **Richard, Philippe**
(1953–   )
French
Physicist

Philippe Richard was born on November 4, 1953, in Montélimar, France, to restaurant owners André Richard and Rose Richard. As a young boy growing up in Montélimar, Richard had interests in astronomy and photography. He attended the Lycée Alain Borne in Montélimar, where he earned his B.S. degree in 1972 in engineer electronics and radiocommunications and an engineering degree in electronics in 1977 from SUPELEC (Ecole Supérieure d'Electricité) in Paris. He then attended Stanford University for graduate studies and received an M.S. in atmospherics and space physics in 1978.

Richard has 20 years' experience in the lightning-detection and thunderstorm-forecasting business. He was first published in 1985 in *Radio Science,* a paper entitled "VHF-UHF interferometric measurements, applications to lightning discharge mapping," which described the interferometric technology design for mapping lightning discharges. Richard started his thunderstorm careers in the late 1970s and early 1980s at the French National Aerospace Research Institute ONERA by diving into the research of the physics of intracloud electricity and lightning strikes. He designed real-time, long-range, interferometric technology for localization and monitoring of total lightning activity of thunderstorms (called SAFIR technology) between 1984 and 1988. Total lightning activity (cloud and cloud-to-ground lightning), produced by thunderstorm cells, provides a new type of information on thunderstorms developments; it can be correlated to severe thunderstorm phenomena such as intense rainfall or downbursts/microbursts and is complementary to other real-time weather data (weather radar, satellite) in improving thunderstorm nowcasting. Nowcasting is short-term weather forecasts, typically for less than the next six hours; in the case of thunderstorms it rarely goes beyond 30 minutes to one hour. The SAFIR systems are networks for lightning localization that have a national, countrywide coverage. The system comprises interferometric VHF detection stations located in various parts of the observed area, connected into a detection network and a central processing system, and is accessible through a PC. The SAFIR system displays the total lightning activity and density on a map of the selected area, and thunderstorm cells and hazard areas can also be tracked. Finally, the user has a nowcast of the thunderstorm as well as automatic warnings available in real time.

All of his near dozen publications deal with lightning research, and he has been recognized for his work by receiving the Blondel Award, French Society of Electrical Engineers (1994); French Air Space Academy "Grand Prix" Award (2000) and Montgolfier Award; and French National Industry Association (2002). In 2002, he married Antje Kruse from Rendsburg, Germany, formerly in international sales.

He continues to make contributions in weather and science by making total lightning-detection technology available to meteorologists. He is currently the general manager of the Thunderstorm Business Unit for Vaisala GAI in Arizona. Vaisala develops, manufactures, and markets electronic measurement systems and equipment for meteorology, environmental sciences, traffic safety, and industry.

⊠ Richardson, Lewis Fry
(1881–1953)
English
Mathematician

Lewis Richardson was born on October 11, 1881, in Newcastle upon Tyne, Northumberland, England, to a family of tanners and Quakers. As a youth, he attended the Newcastle Preparatory School where he studied, among other things, the works of Euclid. From 1894 to 1898, he went to school in York and then spent two years in Newcastle at the Durham College of Science studying mathematics and science. He received a degree in natural science at King's College in Cambridge in 1903. In 1926, Richardson was awarded a Ph.D. in physics from the University of London and was elected as a Fellow of the Royal Society. In 1929, he was given a bachelor's degree in psychology at the University of London.

Richardson first worked at the National Physical Laboratory (NPL, 1903–04), then taught at the University College Aberystwyth (1905–06), and became a chemist for the National Peat Industries (1906–07) before returning to the NPL (1907–09). In 1909, he married Dorothy Garnett, whose father William was assistant to James Clerk MAXWELL and later became head of the physical and chemical laboratory of the Sunbeam Lamp Company (1909–12). Richardson took a position at the Manchester College of Technology (1912–13) and then moved on to the meteorological office at the company's Eskdalemuir Observatory (1913–16) and its office at Benson, Oxfordshire (1919–1920). From 1920 to 1929 he headed the physics department at Westminster Training College, and from 1929 to 1940, he was principal of Paisley College of Technology in Scotland.

Richardson was the first scientist to apply mathematics, in particular the method of finite differences, to predicting the weather. Richardson tried a radically different approach to forecasting, using as his starting point a system of fundamental physical principles governing atmospheric motion. He put together mathematical equations that represented these principles and developed a mathematical method of calculating them. He predicted that if he started from the state of the atmosphere at a particular time—with the initial conditions—his method could be used to work out its future conditions. He carried out two years of manual calculations, much of it in the Champagne district of France, where he served as an ambulance driver for the Friends Ambulance Unit in France as a conscientious objector during World War I (1916–19). However, he used an unrealistic value in the changes in pressure and winds at two locations in central Europe during a six-hour period, and the results were disastrous. Nevertheless, he published the results in his remarkable *Weather Prediction by Numerical Process* in 1922 and is considered the father of numerical weather prediction (NWP). It was not until the 1950s when John von NEUMANN and Jule

Gregory CHARNEY used computers successfully that numerical forecasting became a reality. In retrospect, meteorologists contend that Richardson's math was fine, and much of prediction today is carried out similarly to his; he just had bad data. Today, *Weather Prediction by Numerical Process* is considered one of the most important books in the field.

Richardson proposed a computing theater with 64,000 human computers to forecast the whole planet, but he considered his proposal impractical and called it a fantasy, purely science fiction. In 1975, Nigerian-born Philip Emeagwali, a mathematician and computer scientist, took Richardson's idea and created the 64,000 Hyper-Ball international network of hypercube computers with 64 binary thousand processors to perform the world's fastest computation. He designed his network forecast weather and track global warming, but it was ignored by most scientists.

Another application of mathematics by Richardson, a pacifist, was in his study of the causes of war, no doubt influenced by his experiences as an ambulance driver. In 1919, he self-published *Mathematical Psychology of War*, dedicated to his ambulance colleagues, because he found no publisher willing to print it. He had better luck with *Generalized Foreign Politics* (1939), *Arms and Insecurity* (1949), and *Statistics of Deadly Quarrels* (1950). Unfortunately for him, his conclusions ended up opposite of what he believed—that war was inevitable.

According to Philip J. Davis, professor emeritus of applied mathematics at Brown University: "To numerical analysts, Richardson is known for his algorithms for the numerical solution (in pre-computer days) of ordinary and partial differential equations. The terms jury problem for elliptic equations and marching problems for hyperbolic equations come from him. The Richardson "deferred approach to the limit, [that is, a method for the acceleration of convergence] is fundamental and widely employed."

He made contributions to the calculus and to the theory of diffusion, in particular eddy-diffu-sion in the atmosphere. Atmospheric modelers generally treat atmospheric mixing as a macroscopic eddy-diffusion process. He was not happy about his research on the flow of air being used by the military in the dispersal of poison gases.

The Richardson number, a fundamental quantity involving gradients of temperature and wind velocity, is named for him. In turbulence theory, it characterizes the fraction of turbulent energy that comes from temperature gradients, as opposed to wind velocity gradients.

He produced 117 publications and 10 unpublished works. He died on September 30, 1953, in Kilmun, Argyll, Scotland. In 1972, the Richardson Wing of the headquarters of the British Meteorological Office was opened in his honor. Richardson's many work experiences tossed between chemistry and meteorology, but it was his works in mathematics and pacifism that he has made the biggest impact.

Rossby, Carl-Gustaf Arvid
(1898–1957)
Swedish/American
Meteorologist

Carl-Gustaf Arvid Rossby was born on December 28, 1898, in Stockholm, Sweden, to an engineer and graduated from Stockholm University where he obtained a doctorate in 1925. Rossby was a disciple of the Bergen School of Meteorology founded by Vilhelm BJERKNES.

In 1926, he came to the United States on a Scandinavian–American fellowship to the U.S. Weather Bureau but left Washington on completion to establish the first airways meteorological network in California, funded by Harry Guggenheim. While at the Weather Bureau, Rossby became aquainted with Francis W. Reichelderfer, who also knew Guggenheim, and was a believer in the Bjerknes Bergen School of Meteorology. The weather bureau was reluctant to accept the air-mass method of forecasting that was developed in the Bergen School.

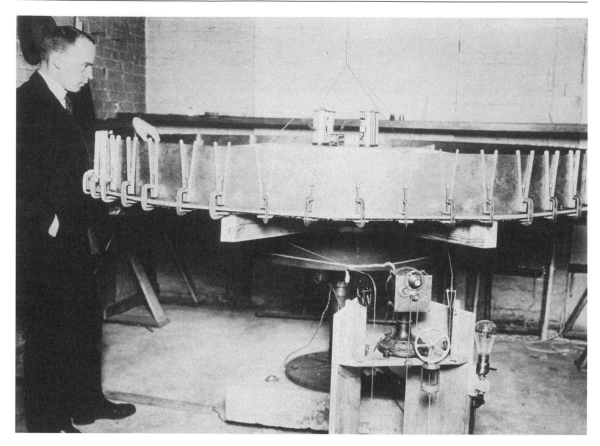

Carl-Gustaf Arvid Rossby. Rossby moved to the University of Chicago (1941–50) and established what is known as the Chicago School of Meteorology. There he developed dynamic models of the general circulation of air. He is considered one of the greatest meteorologists of all time. *(Courtesy of NOAA Image Library)*

Reichelderfer, who played a role in the establishment of the Massachusetts Institute of Technology's meteorology department (also funded by Guggenheim), supported Rossby as its head. Rossby became professor at MIT, founded the study of meteorology and physical oceanography there, and was appointed to the faculty in the Department of Aeronautics in 1928. This later developed into the first department of meteorology in a U.S. academic institution by 1941. He also established collaborations with the new Woods Hole Oceanographic Institution.

In 1939, Rossby became a U.S. citizen and the assistant chief of the U.S. Weather Bureau

(1939–41), trying to redirect scientific efforts to incorporate important Bergen School advances relating to weather fronts and storms. His old friend Reichelderfer became chief of the Weather Bureau in 1938 and asked him to take the job. Both of them professionalized the bureau by making employees take courses and by incorporating the Bergen School science where possible.

Rossby theorized that the Earth's rotation might cause long, slow moving waves in Earth's oceans and atmosphere. He discovered in the 1930s what are now known as Rossby (or planetary) waves, which describe the flow of air within the jet stream (and within oceans), and the

Rossby equation that calculates how fast the flow develops. Rossby waves are generated by the rotation of the Earth, and the waves always travel from east to west, moving very slowly a few miles per day. Rossby named the jet stream during the mid-1940s during the Second World War. Reid A. Bryson and others made calculations that proved the existence of a high-altitude stream of strong winds for a bombing raid on Japan.

In 1991 and 1992, the launches of two satellites, ERS (ESA Remote Sensing) and TOPEX/ POSEIDON, expanded the study of Rossby waves. In 1996, American scientists demonstrated that Rossby waves can be found in all the major ocean basins and that their speed is about twice as fast as had been predicted. Scientists have concluded that this means that the ocean responds twice as quickly to climate change.

Rossby moved on to the University of Chicago (1941–50), and by 1942, he had established what is known as the Chicago School of Meteorology where he developed dynamic models of the general circulation of air. The school also established the importance of vorticity (rotation around a fluid) theories of wave motion in the atmosphere and the ocean. In addition to establishing the Institute of Tropical Meteorology in Puerto Rico, where Herbert Riehl began the development of models of tropical weather disturbances with Joanne SIMPSON, Rossby developed meteorological programs to train cadets for the war. In 1942, he organized and chaired the University Meteorological Committee (UMC), which created 20 premeteorology programs at universities and colleges across the country. In January 1943, the University of Chicago received an average of 2,000 inquiries per day about the program. He was also elected a member of the National Academy of Sciences that same year.

Rossby was a member of the Bergen School under Bjerknes during his early days of study. Norway's Bjerknes, known as the father of numerical weather prediction, theorized that mathematics could be used to predict weather patterns, provided that you had enough information about the state of the atmosphere at a given point in time. Bjerknes theory was correct, but he did not possess the computing power to prove it. It was not until 1954 when John von NEUMANN used the Princeton University's Institute for Advanced Study computers in the Joint Numerical Weather Prediction Unit (JNWPU) Project to prove that numerical weather prediction was feasible. Neumann and Rossby had discussed the prospect for numerical weather forecasting, and Rossby arranged for Jule CHARNEY to assist and head up the meteorology group for Neumann. The method for making the first forecast incorporated ideas developed by Rossby and his colleagues at MIT. The team produced the first computer-generated numerical forecast for weather patterns 24 hours in advance.

Rossby returned to Sweden in 1950 at the request of the Swedish government to help found the Institute of Meteorology at the University of Stockholm. In Stockholm, the Royal Swedish Air Force Weather Service became the first in the world to begin regular real-time numerical weather forecasting, including the broadcast of advanced forecasts. The Rossby's Institute of Meteorology at the University of Stockholm developed the model whereby Forecasts for the North Atlantic region were made three times a week, starting in December 1954. Rossby died on August 19, 1957, in Stockholm.

S

Salm, Jaan
(1937–)
Estonian
Atmospheric Scientist

Jaan Salm was born in the rural district of Kehtna, Estonia, on October 8, 1937, to Johannes and Marie Salm, both farmers. Their peaceful village life was disrupted by the Soviet occupation in 1940 and by the events of World War II. Jaan Salm began his education at a local grade school in 1945 and already had an interest in radio engineering when Estonia was occupied anew by the former Soviet Union. He finished seven grades in six years and continued his studies in the field of radio engineering at the Tallinn Polytechnical School. He finished the polytechnical school with the diploma and honor certificate in 1955. According to the Soviet regulations of the time, graduates of this type of school had to work for a certain period after graduation. Salm's first job was at the Tallinn TV Center, which had just begun to broadcast, and he worked as a technician at the sender station.

In 1956, Salm entered the University of Tartu, Estonia, and majored in physics at the Faculty of Science, graduating with a diploma in theoretical physics in 1960. As an undergraduate, he had worked with a research group in the field of atmospheric ionization and aerosols. After grad-

uation, he was assigned to the position of laboratory assistant at the chair of general physics, University of Tartu, and he continued working with the same research.

The main task of the research group was to design instruments for measuring air ions—electrically charged submicroscopic particles. An ion may be an atom, a molecule, a group of molecules, or a particle (such as dust or a liquid droplet). Both the cluster ions generated by ionizing radiation and the charged ultra fine aerosol particles are considered as air ions. Salm was a creative researcher and soon designed a new type of air ion counter. He was issued a Soviet patent in 1962 for his innovative air ion spectrometer. This patent was also Salm's first scientific publication. A few years later, he received an honor diploma for the air ion counter at an All-Union exhibition.

Inspired by the research of his colleague Hannes Tammet, Salm worked with several fundamental problems in the development of air ion measuring instrumentation. He started his postgraduate study in atmospheric physics at the faculty of science of the University of Tartu in 1964. Primarily, he studied the possibilities for increasing the spectral resolution of air ion spectrometers (a spectroscope equipped with scales for measuring wavelengths or indexes of refraction). The spectral resolution is limited mainly by

Jaan Salm. Salm's most important contribution to meteorology is the characterization of the ultrafine atmospheric aerosols by means of air ion measurement technology. *(Courtesy of Jaan Salm)*

molecular and turbulent diffusion of particles. The molecular diffusion could be considered theoretically. However, complex experimental equipment had to be designed for the study of turbulent diffusion. (Example. If sugar is put in coffee and is not stirred, the sugar mixes with the coffee only by molecular diffusion, very slowly. If the coffee is stirred, turbulence is generated inside it, and turbulent diffusion mixes the sugar more efficiently and quickly than molecular diffusion). Together with his team members, Salm was awarded the Prize of Soviet Estonia in 1967. This prize was the highest scientific award in Soviet Estonia.

Salm successfully defended his thesis "Diffusion Distortions at the Measurements of Air Ion Spectra" at the University of Vilnius, Lithuania, in 1970 and received the diploma and the degree of the candidate of science in geophysics (the degree of the candidate of science awarded in the former Soviet Union is equivalent to a Ph.D.). Salm was hired as a senior lecturer, a full-time teaching position, at the chair of general physics of the University of Tartu in 1969. To further his academic career, Salm was able to attend a half-year of postdoctoral study at the Humboldt University in Berlin in 1973–74. As a nonmember of the Communist Party, Salm had limited opportunities of travel to Western or capitalist countries. However, due to extreme efforts, Salm participated in the XVI General Assembly of IUGG (International Union of Geodesy and Geophysics) in Grenoble in 1975. After several refusals by Soviet officials, he also succeeded in visiting the Paul Sabatier University in Toulouse for two months in 1984. In 1982–85, he was associate professor at the University of Tartu.

Besides teaching, Salm continued intensive research on the design of instrumentation for air ion and aerosol measurements. He was the principal of several research contracts during the 1970s and 1980s, and he designed and built a number of various instruments together with his colleagues; about 10 USSR patents and 80 scientific papers illustrate his work in that period. The design and building of a multichannel automated air ion spectrometer for the Institute of Atmospheric and Ocean Physics of the USSR was one of his most labor-consuming projects in the 1970s. This spectrometer was the forerunner of posterior multichannel electrical aerosol analyzers. These were used in research in the size spectra of atmospheric aerosols and the mobility spectra of air ions.

Beginning in 1985, Salm devoted himself to full-time scientific research at the position of a senior scientist at the University of Tartu. That same year, with Salm's active involvement, researchers began to take regular air ion measurements at a rural site in Estonia, called Tahkuse Observatory. The Tahkuse Observatory yielded statistically weighty long-term data about air ion concentrations in a wide range of mobilities and sizes. Analysis of the data enabled Salm and his coauthors to find various relationships between air ion characteristics and other atmospheric parameters. In general, the air ion mobility spectra give information about ionizing radiation (radiation that has sufficient energy to remove electrons from atoms) and about ultrafine aerosols (those with diameters less than 0.02 μm [micrometer]).

The restoration of Estonia's independence in 1991 led to significant changes in the organization of scientific research. The overall financial shortage also had its effect: the financing of research dropped, and the number of researchers diminished drastically. However, the team of air ion and aerosol research at the University of Tartu survived the difficulties relatively well, and Salm and the team continued their work at the Institute of Environmental Physics at the University of Tartu, established in 1993 as a result of a university reform policy. Salm's longtime colleague Professor Hannes Tammet was the head of the institute. The Estonian Science Foundation played an essential role in supporting the research. Salm was the principal investigator of three grants received from this foundation in the 1990s. Other promising opportunities arose also for participating in scientific conferences over the world.

By 2002, Salm had published nearly 30 scientific papers, based on the measurements at Tahkuse Observatory. Perhaps the most important contribution to atmospheric science made by Salm, together with his coauthors Urmas Hõrrak and Hannes Tammet, is the discovery of the bursts of intermediate ions (between large and small) in certain atmospheric conditions. These bursts are closely related to the generation bursts of aerosol particles observed by many researchers in later times. The generation bursts of aerosol particles became a focus of research in the field of atmospheric aerosols in the early 2000s. The air ion measurement technology is the most effective means for the detection of early stages of the generation of aerosol particles, when the diameters of

particles are still in the nanometer range. This was confirmed by the results of the European Union project BIOFOR in Finland, where Salm participated in 1999. Together with Aare Luts, Salm has also obtained essential results in the mathematical simulation of the evolution of air ions.

Salm married Siiri Koppel, a chemist, in 1970 and they have two children. Salm's scientific publications currently contain 161 items. The majority of his papers are devoted to air ion and aerosol study. Salm's most important contribution to meteorology is the characterization of the ultrafine atmospheric aerosols by means of air ion measurement technology. He continues the research today at the University of Tartu.

⊠ **Schaefer, Vincent Joseph**
(1906–1993)
American
Meteorologist

Vincent Schaefer was born on Independence Day, 1906, in Schenectady, New York, to Peter Aloysius and Rose Agnes (Holtslag) Schaefer. He and his two brothers and two sisters spent much time in the Adirondacks due to the ill health of his mother. At age 16, the young Schaefer dropped out of high school and began to work at the nearby General Electric Company. As a youth, he founded a local tribe of the Lone Scouts and wrote and printed a tribe paper called "Archeological Research." He maintained a lifelong interest in archaeology and natural history.

Though Schaefer had no formal training in any branch of science, he would become one of meteorology's foremost authorities in cloud physics. He completed a four-year course on toolmaking and for three years worked as a journeyman machinist, a tree expert (in Michigan), and an instrument maker at the General Electric Research Laboratory beginning in 1926. Schaefer was granted a one-month leave to accompany Dr. Arthur C. Parker, New York State archaeologist, on an expedition to central New York. During the 1920s and early 1930s, Schaefer pursued his study of natural history and belonged to hiking, archaeology, and outdoor clubs and conducted many outreach programs. A mutual acquaintance at GE introduced Schaefer to Irving LANGMUIR, a brilliant scientist in the research laboratory of the General Electric Company (now the Global Research Center).

In 1931, Langmuir asked the young Schaefer to become his research assistant in the laboratory, along with Katherine Blodgett. The following year, Langmuir won the Nobel Prize in chemistry, and for the next 20 years, this relationship between Langmuir and Schaefer was mutually beneficial as they solved one problem after another. Langmuir would pose the problem, mostly in surface-chemical problems at that time, and Schaefer would devise simple experiments to prove it one way or another. During the 1930s, both published papers on various aspects of surface chemistry. Both, however, had an interest in the mysteries of clouds and the origin of rain and snow.

Since his youth, Schaefer, like Wilson BENTLEY, wanted to study the shape and structure of snowflakes. Unlike Bentley, however, Schaefer did not have the patience for photomicrography, so he invented a method in 1940 to make a perfect replica of a snowflake captured in a thin layer of clear plastic, a technique now used around the world. An outgrowth of this discovery, according to his friend Duncan BLANCHARD, was the first practical method of aluminizing the picture-producing surface of television tubes, more than doubling the contrast and brightness of the picture.

Shortly before World War II broke out, the government asked Langmuir and Schaefer to design a filter for gas masks to trap toxic smoke. This research led to their development of a smoke generator to screen military operations, thousands of which were used before the war ended. He demonstrated the success of his smoke generator at Vrooman's Nose in the Schoharie Valley, a subject later captured in a book he wrote

about the area. It also led to their study of precipitation static on aircraft that interfered with radio transmission. By studying on top of Mount Washington in New Hampshire, Schaefer discovered that clouds were composed mainly of supercooled water droplets. Wanting to know why, Schaefer created his now famous cold-box experiment in his laboratory.

He took an everyday GE home freezer and lined it with black velvet for a dark background. He breathed down into the freezer, at temperatures below 0°C, to produce a supercooled cloud, and with a microscope lamp shining down, he was able to see ice crystals. Although he did not see many ice crystals, he experimented with a variety of materials to convert the water droplets into ice crystals. In July 1946, after noticing that the air in the cold box was not as cold as usual, he dropped a block of dry ice in the box. Within seconds, the entire cloud of water droplets disappeared and turned into tiny ice crystals. After more experimenting, he found that anything cold (about −400°C) introduced into a supercooled cloud will convert it into a cloud of ice crystals. This landmark discovery appeared in *Science* magazine on November 15, 1946; two days before this he performed the first dry-ice seeding of a natural cloud.

On November 13, Schaefer and his pilot Curtis Talbot, in a small single-engine airplane at 14,000 feet, approached a large cloud 30 miles east of Schenectady, New York. Irving Langmuir was on the ground watching through binoculars. When they flew into the cloud, Schaefer dropped 3 pounds of dry ice, and, within seconds, long streams of snow began to fall from the base of the cloud. It was the first successful demonstration that a natural supercooled cloud could be converted at will into a cloud of ice crystals. As Duncan Blanchard writes, "The modern science of cloud physics and experimental meteorology had begun." The following day, Bernard VONNEGUT discovered, also in a cold box, that silver iodide was an effective seeding agent.

In the spring of 1947, the five-year government-sponsored Project Cirrus began, and Langmuir, Schaefer, Vonnegut, and others made important discoveries, including the invention of new instruments and the establishment of practical seeding technology. It was during Project Cirrus that Vonnegut first used silver iodide to seed natural supercooled clouds. In 1954, Project Cirrus closed and Schaefer left GE to become director of research of the Munitalp Foundation where he initiated many cooperative research programs, including one on orographic (the study of the physical geography of mountains and mountain ranges) and noctilucent (luminous at night) clouds with the International Institute of Meteorology in Stockholm, a worldwide program of time-lapse photography, and a lightning research program (Project Skyfire) with the U.S. Forest Service. Schaefer left Munitalp in 1958, turning down an offer to move with the Foundation to Kenya, but he remained an advisor to Munitalp for several years.

In the 1960s, he was the prime mover in establishing the Atmospheric Science Research Center (ASRC) at the State University of New York at Albany and became the first director of research and then overall director. It was at this time that Vonnegut, Raymond Falconer, and Blanchard, three members of Project Cirrus, moved to Albany to work with Schaefer. Schaefer also hired Roger CHENG to run his laboratory shortly after. ASRC is a nationally recognized research center and continues to operate out of the University of Albany.

Schaefer always knew the value of education though he never finished high school. In 1959, he created the National Sciences Institute and through a decade worked with more than 500 students from around the United States as they took part in his outdoor adventures. From 1959 to 1961, Schaefer was director of the Atmospheric Science Center at the Loomis School in Connecticut. During the winters of 1959 to 1971, Schaefer led weeklong excursions into the Old Faithful area of Yellowstone National Park on natural historic expeditions for scientists.

During his career, he wrote more than 250 papers covering a wide range of topics including natural history, Dutch barns, archaeology, and even how to increase heat dissipation from a fireplace. He retired from ASRC in 1976 at age 70 and continued his interests, particularly on the preservation of local Dutch barns with the Dutch Barn Preservation Society. In his retirement, Schaefer collaborated with photographer and cloud physicist Dr. John DAY on the popular *A Field Guide to the Atmosphere* (1981).

A man who did not finish high school, Schaefer received honorary doctorates from University of Notre Dame (1948), Siena College (1975), and York University (1983). He was a Fellow of the Rochester Museum of Arts and Sciences (1943) and the American Association for the Advancement of Science (1956). He also was awarded numerous honors, including Schenectady Junior Chamber of Commerce Award; Young Man of the Year (1940); CSIRO, Australia (1960); and American Meteorological Society (1967). He also received awards from the American Geophysical Union, First Paper of Outstanding Excellence (1948); Man of the Year Award, Notre Dame Club of Schenectady (1948); Member (Hon.) Union Chapter, Sigma Xi; Institute of Aeronautical Sciences, Robert M. Losey Award (1953); The American Meteorological Society, Advancement of Applied Meteorology Award (1957); Fellowship, Woods Hole Oceanographic Institution (1959); Member (Hon.) Sigma Pi Sigma, Albany Chapter (1960); Distinguished Science Lectureship, State University of New York, College of Education at Albany (1960–61); Ideal Citizen of the Age of Enlightenment Award—All Possibilities, Research and Development Award, American Foundation for the Science of Creative Intelligence (1976); Weather Modification Association Vincent J. Schaefer Award for scientific and technical discoveries that have constituted a major contribution to the advancement of weather modification (1976); and Citizen Laureate, University at Albany Foundation (1980).

Schaefer died on July 25, 1993, at the age of 87. His long time assistant Roger Cheng has prepared a CD-ROM collection of his work and career.

⊠ **Sellers, Piers John**
(1955–)
English/American
Climatologist

Piers Sellers was born on April 11, 1955, in Crowborough, United Kingdom, one of five sons born to John Alexander Sellers, a British army officer, and Hope Lindsay Sellers. As the Sellers family was constantly moving from one military posting to another, Piers was sent to boarding schools from the age of seven. Preparatory schools included Tyttenhanger Lodge and Glengorse in Sussex (age 7–13), and Grammar school (age 13–18) at Cranbrook School in Kent. As a youth, he was interested in science and flying, and at the age of 16, he had a glider license and earned his pilot's license the following year. He also had an interest in space. After watching Neil Armstrong step on the moon, he decided that he wanted to become an astronaut.

He attended the University of Edinburgh (1973–76), receiving a bachelor of science degree (with honors) in ecological science, and Leeds University (1976–81), receiving a Ph.D. in climatology. In 1980 he married Amanda Helen Lomas. They have two children.

Sellers began his career as a programmer/systems analyst for Scicon, a computer consulting firm in London (1981–82), where he worked on geophysics exploration software for BP Minerals PLC. From 1982 to 1984, he was a sponsored resident research associate for the U.S. National Research Council, working on surface-energy balance computer models at NASA's Goddard

Piers John Sellers. Sellers has worked on research into how the Earth's biosphere and atmosphere interact, using computer modeling of the climate system, satellite remote sensing studies, and coordinated fieldwork utilizing aircraft, satellites, and ground teams in a variety of places, including Kansas, Russia, Africa, Canada, and Brazil. *(Courtesy of Piers John Sellers)*

Space Flight Center, along with Yale MINTZ. He left to become a faculty assistant research scientist for the department of meteorology at the University of Maryland until 1990. That year, he began to work for NASA as staff scientist, and in 1996, he accomplished his boyhood dream—he became an astronaut. He is now a mission specialist at NASA's Johnson Space Center.

Sellers has worked on research into how the Earth's biosphere and atmosphere interact, using computer modeling of the climate system; satellite remote-sensing studies; and coordinated fieldwork utilizing aircraft, satellites, and ground teams in a variety of places, including Kansas, Russia, Africa, Canada, and Brazil. He has logged more than 1,100 hours as a pilot of various light aircraft (singles and twins) that have been used in these experiments. Most of this work was done while he was based at or near NASA's Goddard Space Flight Center.

Sellers has made a number of contributions helping to explain how vegetation interacts with the atmosphere. He was one of the founding members of the scientific team that developed and put into practice to this end a series of integrated large-scale field experiments, such as the First ISLSCP (International Satellite Land Surface Climatology Project Field Experiment (FIFE) and the Boreal Ecosystem-Atmosphere Study (BOREAS). These experiments were designed to validate land surface–atmosphere models, using surface and airborne observations, and to develop techniques for adapting these models for use in atmospheric general circulation models (AGCMs) by using satellite data as an integration tool. Sellers expanded this work to include study of surface–atmosphere carbon exchange, in addition to radiation, heat, and water fluxes. He also worked on the development of analyses, tying together vegetation canopy function (photosynthesis, transpiration), radiative transfer, and remote sensing with the use of this methodology to describe vegetation function inside AGCMs using satellite data. An AGCM calibrated in this way was used to calculate the effect of doubled CO_2 on global vegetation and its feedback onto the atmosphere. In the calculation, vegetation transpired less water due to its exposure to increased CO_2, and continental heating was additionally augmented over previous greenhouse calculations.

Sellers has authored more than 150 publications in journals, books, and proceedings and has chaired or managed several important committees and projects. He has also been the recipient of a

number of awards and honors, including the American Meteorological Society's Henry G. Houghton Award for "outstanding achievements in the development and field testing of models describing land–atmosphere interactions" (1997); several NASA awards (1990; 1992–96); the Arthur S. Fleming Award (1995); and American Institute of Aeronautics and Astronautics (AIAA) Award for advancing professional growth in understanding new technology in the field of atmospheric sciences (1990). He was elected a Fellow of the American Geophysical Union in 1996 and a Fellow of the American Meteorological Society in 1998.

Sellers is an astronaut (spacewalker) at NASA's Johnson Space Center, training for a launch to the *Space Station*. He continues to be closely interested in studies of the global biosphere.

Sentman, Davis Daniel
(1945–)
American
Physicist

Davis Sentman was born on January 19, 1945, in Iowa City, Iowa, where he grew up on an Iowa farm but always had an interest in science. He attended Ollie Community School for his early education and then Pekin Community High School, graduating in 1963. He worked as a standards checker at the Swift Feed Mill in Des Moines, Iowa, in 1964. From 1965 to 1969, he served in the air force as a radar technician/technical control specialist and then earned a B.A. in mathematics in 1971 and an M.A. in physics from the University of Iowa in 1973. That same year he served in the Peace Corps for several months in Kenya as a physics and mathematics teacher and then returned to the University of Iowa, receiving his Ph.D. in physics in 1976 with his thesis entitled "Whistler Mode Noise in Jupiter's Magnetosphere." The thesis used energetic particle data from the 1973 *Pioneer 10* spacecraft encounter with Jupiter to predict what the levels of electromagnetic radiation would be in the radiation belts. Subsequent observations by Voyager spacecraft in 1977 generally confirmed these predictions.

Sentman began to publish while in graduate school in 1973 when his research on solar physics resulted in his article "A Search for 5 Min Periodic Structure in Solar 2-Cm Emission" published in *Solar Physics*. This work investigated whether the well-known 5-min oscillations of the solar atmosphere could be detected in microwave emissions. They could not.

Sentman has made three important contributions to solar-terrestrial research. He determined that a class of plasma waves called odd half harmonic emissions found in the plasmasphere that had been previously thought to be due to instabilities was, in fact, a type of stable electrostatic thermal fluctuations. The plasmasphere is thought to be created by sunlight-energizing molecules in Earth's upper atmosphere and is contained by Earth's magnetic field. He has also been at the forefront of research into lightning effects extending into the middle and upper atmosphere, coining the term *sprites* for the phenomenon. This is a new area of geophysical research, and current work by Sentman and others is focused on investigating previously unknown electrical linkages between the thunderstorms and upper atmosphere, as well as possible associated chemical effects. Finally, he has been actively involved in recent experiments creating artificial airglow in the upper atmosphere using high-power radio transmitters located in Alaska. This work provides investigators with a controlled laboratory and new capabilities for studying microphysical plasma and photochemical processes in the upper atmosphere.

Sentman has received a number of awards, such as the Laurel Award from *Aviation Week & Space Technology Week* for aircraft operations during thunderstorm investigations (1994); the Terris and Katrina Moore Prize, University of Alaska

(1995); and, in 1997, the Emil Usabelli Award, University of Alaska. He has published more than 60 articles and research reports, given numerous talks, and served as a reviewer for several journals and funding sources. Sentman has served on a variety of committees and panels and has been principal investigator on 20 NASA, National Science Foundation, Department of Defense, and Department of Energy grants and contracts. He has been the special editor for the *Journal of Atmospheric and Solar-Terrestrial Physics* since 1996.

In 1991, Sentman became an associate professor in the physics department at the University of Alaska, where he continues to teach, becoming full professor in 1996, and to conduct his research on extending studies of meteorological processes into the upper atmosphere/ionosphere/magnetosphere. He also serves as mentor and advisor to several students who are working on their doctoral theses.

Shaw, Sir William Napier
(1854–1945)
English
Physicist, Meteorologist

Napier Shaw was born on March 4, 1854, in Birmingham, West Midlands, England. After studying at Cambridge, he worked at the Cavendish Laboratory for experimental physics from 1877–1900 under James Clerk MAXWELL, Lord RAYLEIGH, and J. J. Thompson. Rayleigh appointed Shaw as demonstrator at the lab.

In 1905, he became director of the U.K. Meteorological Office. The following year, he and Rudolf Gustav Karl Lempfert helped to establish the polar front theory of cyclones, established later by Vilhelm BJERKNES and the Bergen School. This was first published in 1906 in a paper "The life history of surface air currents: a study of the surface trajectories of moving air" and was followed by a book the same year, coauthored by Shaw and R. G. K. Lempfert, called *Life History of Surface Air Currents*.

He also introduced the millibar as the meteorological unit of atmospheric pressure in 1909; it would take another decade before it was used internationally. Following the Eighth Congress of the World Meteorological Organization, starting on January 1, 1986, the term *hectopascal* (hPa) became preferred to the numerically identical *millibar* (mbar) for meteorological purposes.

In 1914, Shaw defined geostrophic winds, which are winds, about 3,300 feet above the ground, that are largely driven by temperature (pressure) differences and are not influenced greatly by the Earth's surface. The following year, he was knighted.

Meteorologists utilize thermodynamic, adiabatic, or aerological charts or diagrams to study certain weather conditions, and all of them contain five elements: isobars, isotherms, dry adiabats, pseudo-adiabats, and saturation moisture lines. Shaw is credited for inventing the tephigram, a thermodynamic diagram widely used in the field. The tephigram determines atmospheric stability and moisture and the atmosphere's vertical structure. The tephigram is named from combining the two elements used in its preparation, temperature and entropy: *TE* stands for temperature, and the Greek letter phi was used to denote entropy. It is commonly used today by aviation meteorologists to display a vertical profile of the atmosphere. The other two diagrams are called the SkewT/Log P diagram and the Psuedoadiabatic (or Stüve) diagram.

These charts are prepared in color to differentiate the various components. Because Shaw was director of the Met Office from 1905 to 1920, his diagram was accepted and used by the Met office and affiliates, being officially adopted by the International Commission for the Exploration of the Upper Air in 1925. Shaw also was a pioneer in the study of the upper atmosphere by deploying meteorological instruments carried by kites and high-altitude balloons.

Shaw was the President of Royal Meteorological Society in 1918–19 and published a number of articles and books which he wrote with his colleagues Lempfert and J. S. Owens, including *Forecasting Weather Meteorology* (1913), *The Air and its Ways* (1923), *The Smoke Problem of Great Cities* (1925) (a pioneering work on atmospheric pollution), *The Drama of Weather* (1934), and the four-volume set *Manual of Meteorology* (1926–31).

He died on March 23, 1945.

Simpson, Joanne Gerould
(1923–)
American
Meteorologist

Joanne Gerould Simpson was born on March 23, 1923, in Boston, Massachusetts, the daughter of Russell Gerould, a reporter for the Boston Herald. Her mother, Virginia (Vaughan) Gerould, was also a former reporter. Simpson attended a private high school, graduating in 1940, but at the age of 16, she had her own pilot's license, a talent she would later use successfully in meteorology. Her talent for physics and mathematics was evident while in school. She went on to the University of Chicago in Illinois where she developed an interest in meteorology, receiving a bachelor's degree in 1943 and a masters degree (in meteorology) in 1945. She studied there with the 20th century's most famous meteorologist, Carl-Gustaf ROSSBY.

During World War II, Simpson was hired by the military to teach aviation cadets at New York University and students at Chicago the basics of meteorology so they could act as weather forecasters in the field. However, women were supposed to go home after the war, and as she wrote in a self-reflection piece, "Like Rosie the Riveter, we were supposed to go home and mind the baby and the mop"; instead, she returned to the University of Chicago after the war. In spring 1947, she took a course on tropical meteorology with Professor Herbert Riehl, who became her mentor for her Ph.D. program after being thoroughly rejected by ROSSBY as a waste of time, citing a male dominated field. She received a Ph.D. in meteorology in 1949, the first woman to do so and became the first professional women meteorologist in the world.

She accepted a teaching position at the Illinois Institute of Technology as an assistant professor in Chicago, from 1949 to 1951. From there, Simpson went to Woods Hole Oceanographic Institution in Woods Hole, Massachusetts, and studied weather patterns over oceans until 1961. At Woods Hole, she used her piloting experience by flying a surplus WWII PBY Catalina flying boat to study clouds.

She continued to collaborate with Herbert Riehl, developing the "Hot Tower Theory," which states that deep convective clouds in the equatorial region were responsible for transporting heat to the upper troposphere. They collaborated on tropical meteorology, on the role of convective clouds in tropical circulations, and on hurricanes until his death. Simpson went on to become a professor of meteorology at the University of California at Los Angeles from 1961 to 1965. A year after entering the University at California, she was awarded the American Meteorological Society Meisinger Award and was elected as a Fellow in 1968. This research led her to be selected as an advisor to the Hurricane Research Program. She married the director of the project, Robert SIMPSON, a meteorologist from the U.S. Weather Bureau, on January 6, 1965, and combined the families of their previous marriages; she had three children from a previous marriage, and he had two.

This position was followed by a stint with the National Oceanic and Atmospheric Administration's Experimental Meteorology Lab in Coral Gables, Florida, until 1974. There, Simpson explored techniques of cloud seeding into hurricanes, which was thought to increase rainfall and possibly weaken the fury of hurricanes. Cloud seeding had been successfully developed by SCHAEFER and VONNEGUT 30 years prior.

The idea was tested in 1961, 400 miles north of Puerto Rico, in the eye wall of Hurricane Esther. The hurricane stopped growing and even showed signs of weakening, helping establish Project Stormfury. The next hurricane, Hurricane Beulah, was seeded on August 23, 1963, and it too showed signs of weakening. However, it was not until 1969 that another hurricane, Debbie, was seeded five times in August. The wind speed decreased about 30 percent and after a second seeding reduced to 15 percent. However, further research has shown that hurricanes weaken and grow naturally, so by 1980, Project Stormfury ended without proving that hurricanes could be weakened using those methods.

From 1974 to 1979, Simpson was professor of environmental sciences at the University of Virginia, in Charlottesville, Virginia, and from 1979 to 1988, she served as the head of the Severe Storms Branch of NASA's Goddard Space Flight Center (GSFC) in Greenbelt, Maryland. In 1983, she received the Carl-Gustaf Rossby Research Award, one of the most prestigious prizes in her field and the American Meteorological Society's highest honor, named for her former professor who warned her about the difficulties she would face as a woman scientist.

In 1988, Simpson became the chief scientist for meteorology and a Goddard Senior Fellow at the Earth Sciences Directorate of the Goddard Space Flight Center, where she still works today. She was also elected a member of the National Academy of Engineering "for far-reaching advances in the mechanisms of atmospheric convection, clouds, and precipitation and their application to weather prediction and modification."

Simpson became the first—and, to date, only—female president of the American Meteorological Society in 1989. She also held two terms as their councilor in the 1970s, Commissioner of Scientific and Technological Activities between 1981 and 1987, and publications commissioner from 1992 to the present. She was a Fellow of the American Geophysical Union in 1994.

One of the most important projects of her career was as Project Scientist for the joint U.S.-Japan Tropical Rainfall Measuring Mission (TRMM) from 1986 to 1997. A satellite launched in November 28, 1997, was the first spacecraft dedicated to measuring tropical and subtropical rainfall through microwave and visible infrared sensors. The satellite makes it possible for quantitative measurements of tropical rainfall to be obtained on a continuing basis over the entire global tropics. The satellite is still flying in 2003, and she is still analyzing the data. Also on May 14, 1997, NASA's then fastest computer the CRAY T3E was named in her honor, the jsimpson.

Simpson was a pioneer in cloud modeling, producing the first one-dimensional model and the first cumulus model on a computer. In 1957, the last year of Rossby's life, he encouraged her to put her "rudimentary 2D cloud model on his computer in Stockholm." She regards him to this day "as a great positive influence on my life and work."

The Ms. Foundation for Women named her one of its top 10 female role models of 1998. The following year, she was elected an Honorary Member of the Royal Meteorological Society of the U.K. for her valuable contributions to the field of meteorology. Honorary membership is the highest honor bestowed by the Royal Meteorological Society.

Simpson's research has given her recognition as one of the world's leading authorities on cumulus clouds, severe thunderstorms, and hurricanes. She has now spent 60 years studying tropical storms, rainfall, and clouds—especially cumulonimbus clouds. She is still working at Goddard and stated in a newspaper article in 2000 that she has no plans to retire. Her credits include more than 170 publications in the areas of tropical meteorology, tropical cloud systems and modeling, tropical storms, and tropical rain measurement from space.

Simpson is a member of Phi Beta Kappa and Sigma Xi and throughout her career has been awarded many other accolades, such as the Guggenheim Fellowship in 1954 and the C.F. Brooks Award in 1992. She also received the Department of Commerce Gold Medal in 1972, the Professional Achievement Award of the University of Chicago Alumni in 1975 and 1992, the NASA Exceptional Scientific Achievement Award in 1982, and the Women in Science and Engineering (WISE) Lifetime Achievement Award in 1990. Simpson has been listed in *Who's Who of American Women*, 7th Edition, since 1972 and in *Who's Who in America* since 1980. In 2001, she won the Charles Anderson Award of the American Meteorological Society for promoting diversity in the workplace. She has worked hard throughout her life to mentor and encourage women and minorities in earth sciences.

⊠ Simpson, Robert H.
(1912–   )
American
Meteorologist

Robert Simpson was born in Corpus Christi, Texas, on November 11, 1912, and is the oldest of three children to Clyde Robert Simpson, a schoolteacher turned agricultural-hardware business owner. His earliest experience in 1919 with the weather was a frightening one in which he and his family had to swim to safety for three blocks (he on his father's back) in near-hurricane force winds. During his high school days at Corpus Christi, he excelled at learning how to play the trumpet and enjoyed architectural and engineering drawing. On graduation, he became an apprentice architect in San Antonio and designed 30 large family houses, enjoying what looked like a promising career. However, the economic crash of 1929 demolished his dreams of pursuing the field. Yet, it was his other high school talent, the trumpet, that was responsible for Simpson's admission to college. For two years in a row, he won first place in the state trumpet solo contest and was invited to enter Southwestern University of Georgetown, Texas, to be the first-chair trumpet in the orchestra and band. He accepted admission but majored in physics and mathematics instead of music.

Simpson graduated Southwestern University in 1932 with a bachelor of science degree in physics and mathematics and began graduate school at Emory University, where he obtained a master's degree in physics and mathematics three years later. Unable to find a job in the field, he went back to his home roots and began to teach music, eventually taking charge of the Corpus Christi school system's instrumental music program. In 1939, he accepted a position with the National Weather Service in Brownsville and began his meteorological career.

Simpson ended up with more than 55 years experience as an atmospheric scientist, almost equally divided among research, operations, and management and completed a doctorate in geophysical sciences at the University of Chicago in 1962. He served 34 years with the National Weather Service in a wide range of positions, including four years as founding director of the National Hurricane Research Project; seven years as director, National Hurricane Center; three years as Weather Bureau deputy director of research (Severe Storms); and four years as National Weather Service director of operations. In 1974, he founded Simpson Weather Associates of Charlottesville, Virginia, and is presently president emeritus.

Simpson's broadest experience is in the field of tropical meteorology and hurricanes, both as a forecaster and in hurricane research, having planned and conducted dozens of pioneering research flights through hurricanes and other severe storms. Following the initial exploratory flight into a Gulf of Mexico hurricane by Col. Joseph Duckworth in 1943, the first operational hurricane flights of the air force and navy in 1944, and an exploratory flight by Dr. Harry Wexler in an air-force plane in 1944, he planned

and conducted one of the first complete research missions through the eye of a hurricane in August 1945, and in 1947, the first research overflight of a major hurricane.

As director of the National Hurricane Center, and in collaboration with Herbert Saffir, Simpson helped design and inaugurate the use of the Saffir–Simpson damage-potential scale in 1973, used to specify the expected damage from a hurricane. This scale presently is used internationally to define and compare the strength of hurricanes at sea and during landfall.

After leaving government service in 1974, Simpson, with his wife and colleague, distinguished tropical meteorologist Joanne SIMPSON, spent five years at the University of Virginia on the faculty of environmental sciences, during which they were both active participants in international research expeditions, including the GATE (tropical Atlantic), MONEX (South China Sea), ITEX (Timor Sea) programs, and Toga Coare (Maritime Continent). Simpson has lectured at many universities and research institutes in the United States, Canada, Australia, Europe, and Asia. During the last 25 years, he has consulted broadly on problems of the coastal zone and its management.

Simpson has published more than 65 papers in professional (refereed) scientific journals here and abroad and in 1981 published the textbook *The Hurricane and Its Impact* (senior author with Herbert Riehl). He is a Fellow of the American Meteorological Society and a certified consulting meteorologist; also he is a Fellow of the Explorers Club of New York. For his contributions to the knowledge and the prediction of severe storms, Simpson received the Department of Commerce Gold Medal Award (1962). He also holds an honorary doctorate in 1963 from his early alma mater, Southwestern University of Texas, the Gold Medal of France (1973), and the 1990 American Meteorological Society's Cleveland Abbe Award, the citation of which reads "for pioneering work in storm research and for outstanding leadership in planning and implementing complex operational programs over a span of decades." He presently is retired and remains active in professional groups, including The Explorers Club.

Sinkevich, Andrei
(1951–   )
Russian
Cloud Physicist

Andrei Sinkevich was born on December 8, 1951, in Lvov in the former USSR, to Alexandr Leonidovich Sinkevich and Emeljanova Ludmila Andreevna, both scientists and teachers at the Forest Academy in St. Petersburg, Russia. As a boy, Andrei was interested in sports (football, basketball, hockey) and mathematics while he attended school from 1960 to 1969 in St. Petersburg.

In 1975, he received a degree in radio engineering after graduating from Leningrad Institute of Constructing Aircraft Devices. His Ph.D. research in 1982 was devoted to the problem of in-cloud temperature measurement with the help of IR (Infrared) Radiometers. His doctoral thesis in 1992 was devoted to constructing a complex of aircraft meteorological devices and carrying out investigations of clouds northwest of Russia.

Although his first area of work was in engineering, he has made important discoveries in cloud thermal structure with practical scientific application from experiments to prevent precipitation in St. Petersburg during holidays. The study of cloud thermal structure is the basis to understand the rules of development of clouds, and cloud-seeding experiments to prevent precipitation is a rather new field in cloud-seeding physics.

Sinkevich has published 12 scientific papers in leading journals and, in 2002, received the title of honored meteorologist in Russia. He married Belova Tatiana Mishailovna in 1974. They have one son.

Presently the chief of the cloud-physics, cloud-seeding, and solar-radiation-studies department at the A. I. Voeikov Main Geophysical Observatory in St. Petersburg, Sinkevich is currently working on several projects that deal with problems of cloud physics and cloud seeding.

Solomon, Susan
(1956–   )
American
Chemist

Susan Solomon was born on January 19, 1956, in Chicago, Illinois, to Leonard, an insurance salesman, and Alice (née Rutman), a fourth-grade teacher. Solomon was led into the field of science through her interest in the work of Jacques-Yves Cousteau when she was 10 years old. Her first taste of success was winning third prize in the international student science fair with a project on the use of light to determine the percentage of oxygen in gaseous mixture.

Solomon attended the Illinois Institute of Technology at Chicago and earned a bachelor's degree in chemistry with high honors in 1977. In 1977, Solomon received a University Corporation for Atmospheric Research (UCAR) student fellowship and worked at The National Center for Atmospheric Research (NCAR) the summer before she began graduate school. After a few months working under Paul Crutzen, then director of NCAR's Atmospheric Chemistry Division, and with Ray Roble, she was hooked on atmospheric science as a career. She returned to NCAR to do her dissertation under the NCAR graduate assistant program during 1979–81. Solomon attended graduate school at the University of California at Berkeley and received her master's degree and a Ph.D. in chemistry in 1979 and 1981, respectively. While at NCAR, she began to work with Rolando Garcia to develop a coupled, two-dimensional chemical-dynamical model of the stratosphere and the mesosphere. It quickly led to

Susan Solomon. Solomon is the lead or coauthor of 150 scientific papers that have provided key measurements and theoretical understanding regarding Arctic and Antarctic ozone destruction. *(Courtesy of Susan Solomon)*

a better physical understanding of how stratospheric wind moves around such trace chemicals as stratospheric methane and ozone. When the ozone hole was discovered in 1985, Solomon and Garcia were fortunate in having such a good model to examine possible explanations for its mysterious occurrence and found heterogeneous chemistry (in particular, the reaction of hydrochloric acid with chlorine nitrate) as the likely cause. Solomon says this turned out to be a good guess, and it is now widely acknowledged as the key initiating step in producing the ozone hole.

In the mid-1980s, she worked with Jeff Kiehl of the Climate and Global Dynamics Division and began to put a better treatment of radiation into their two-dimensional stratospheric model. In 1986 and 1987, Solomon served as the head project scientist of the National Ozone Expedition at McMurdo Station, Antarctica, where she made some of the first measurements showing that chlorofluorocarbon chemistry indeed is responsible for the ozone hole.

Since then, she has combined this dynamical and chemical knowledge to look at the effects of volcanic eruptions on ozone depletion, at gravity waves and mesospheric species, and at a number of other intriguing chemical-dynamical problems.

Solomon is the recipient of more than 20 awards for her scientific work, including the J. B. MacElwane Award of the American Geophysical

Union (1985); the Department of Commerce Gold Medal for Exceptional Service (1989); the Henry G. Houghton award (1991) and the Carl-Gustaf Rossby Research Medal (2000) of the American Meteorological Society. In 1992, *R&D* magazine honored her as scientist of the year. She is presently a senior scientist at NOAA and head of the Aeronomy Laboratory's Middle Atmosphere group in Boulder, Colorado. In 2000, she received the National Medal of Science, the highest scientific honor in the United States, for "'key scientific insights in explaining the cause of the Antarctic ozone hole."

Solomon is a member of the National Academy of Sciences, a foreign member of the French and European Academies of Sciences, and a Fellow of both the American Meteorological Society and the American Geophysical Union. She has received honorary degrees from University of Colorado (1993), Tulane University (1994), and Williams College (1996). In Antarctica, Solomon Glacier and Solomon Saddle is named in her honor. In 2002, she was awarded the Weizmann Women & Science Award, given biennially to an outstanding woman scientist in the United States who has made a significant contribution through research in basic or applied science.

Solomon is widely known for her crucial role in efforts to determine the cause of the Antarctic ozone hole and showed how chlorofluorocarbons (CFCs) interact in the unique Antarctic environment to cause ozone depletion there. She is currently focusing on research in many fascinating areas, including photochemistry and transport processes in the stratosphere and troposphere; remote sensing of the atmosphere by spectroscopic methods and their interpretation; interpretation of ozone depletion at midlatitudes and in polar regions; coupling between trace gases and the Earth's climate system; and, stratospheric chemistry, especially observations and interpretation of the chemistry of the Antarctic ozone hole.

Her book, *The Coldest March*, cites new data to argue that bad weather, not poor planning, was the reason that Capt. Robert Falcon Scott and his team of British explorers perished in Antarctica in 1912.

Solomon is the lead or coauthor of 150 scientific papers that have provided key measurements and theoretical understanding regarding Arctic and Antarctic ozone destruction. She has been a member of numerous advisory committees for the National Academy of Sciences, the National Science Foundation, and the National Aeronautics and Space Administration. She has testified on the ozone issue to both Senate and House subcommittees concerned with the environment and is widely considered one of the world's leading experts on this topic.

⊠ **Stensrud, David J.**
(1961–   )
American
Meteorologist

David Stensrud was born in Minneapolis, Minnesota, to Stanton Stensrud, an insurance adjuster, and Betty Stensrud, an elementary school teacher. While attending Fair Elementary, Carl Sandberg Junior High, and Neal Armstrong High Schools, he enjoyed the outdoors and sports and was active in band, church choir, and youth group. His lifelong love of the outdoors and the fascination of watching evening thunderstorms in Minnesota helped spur his early interest in the weather.

He attended the University of Wisconsin at Madison where he initially intended to major in physics. By his sophomore year, he realized that he had little passion for high-energy physics and took an introductory class in meteorology that was taught by Frank Sechrist. He was hooked. Meteorology had not only a strong physics component to Stensrud but also a clear public-service mission that was very compelling. Public safety and health, environmental issues from air quality to energy conservation to global climate change,

David J. Stensrud. He documented and explained the dramatic upscale influences of mesoscale convective systems (organized regions of thunderstorms often associated with a trailing region of light to moderate precipitation) during the warm season over midlatitudes. *(Courtesy of David J. Stensrud)*

and many other sciences are all connected to meteorology. He received his bachelor's degree in 1983 with a double major in mathematics and meteorology and then spent the summer as an intern at the Western Region Headquarters of the National Weather Service. He learned a great deal about how to apply his weather knowledge and how to write computer programs during his brief 10 weeks in Salt Lake City. He also gained confidence in his skills and abilities to interact professionally with other meteorologists. Glenn Rasch, head of the Scientific Services Division, was an excellent mentor and allowed Stensrud to take on several forecast-related responsibilities. This internship helped begin many long-lived professional friendships.

Immediately after this summer internship ended, Stensrud began graduate school at the Pennsylvania State University. He obtained a master's degree in meteorology in 1985, the same year he married Audrey Snyder. They have two children. In 1986, they moved to Norman, Oklahoma, for him to begin work as a research meteor-

ologist at the National Oceanic and Atmospheric Administration's (NOAA's) National Severe Storms Laboratory (NSSL). This job was a turning point in his professional career, the place where he finally found his calling—improving numerical forecasts of severe weather events.

His first publication with colleague and adviser H. N. Shirer appeared in the *Journal of the Atmospheric Sciences* in 1988. Entitled "Development of boundary layer rolls from dynamic instabilities," this paper explores the role of wind shear in the development of organized upward/downward circulations within the first 1 to 2 kilometers above the ocean surface, using special data sets obtained by aircraft over the North Sea. Their results suggest that wind-shear instability helped develop these circulations on several of the days that were studied and indicate that simple models can be used to help predict these phenomena.

He received a Ph.D. in meteorology in 1992 from Penn State. Stensrud has made two important contributions in meteorology so far. He documented and explained the dramatic upscale influences of mesoscale convective systems (organized regions of thunderstorms often associated with a trailing region of light to moderate precipitation) during the warm season over the midlatitudes. These systems can easily perturb the large-scale flow patterns over one-quarter of the hemisphere and are an important and challenging forecast concern.

He also has contributed to a better understanding and acceptance of the importance of model diversity to ensemble forecasting. Ensembles are groups of model forecasts that are valid during the same time period. When ensembles were first used in the medium-range forecast problem (7–14 days), the ensemble approaches all used the same forecast model and used variations in the initial atmospheric state to create the different ensemble members. His work on short-range (0- to 48-hour) ensembles clearly indicates the value of including model diversity—that is, using several different models—in an ensemble system. This is in part because in the short range we are interested in weather events that influence people, such as rainfall and high temperatures, and predictions of these variables are very sensitive to how the forecast model is constructed. Recent research by others has shown that model differences play a role in medium-range ensemble forecast systems as well.

Stensrud's work reminds us of how difficult it is to make a good forecast of the weather. The complexities of the atmosphere are enormous, and the interconnections range among the ocean, soil, vegetation, and biosphere. Thus, the faster computers and the sophisticated forecast models available today do not necessarily translate into better forecasts for use by the public. Chaos—the sensitive dependence upon initial conditions—occurs, and this limits the accuracy of predictions. Stensrud stresses that we need to be more creative in our use of tools and data if we expect to produce better forecasts.

In 1996, Stensrud became an adjunct assistant professor at the University of Oklahoma, and since 1999, he has been an adjunct associate professor. He regularly teaches a graduate level computer-modeling course and advises several graduate students, while still conducting his own research at NSSL. He is a member of the Chi Epsilon Pi, a meteorology honorary society, and Sigma Xi, the Scientific Research Society. In 1996, he was given the White House Presidential Early Career Award for Scientists and Engineers by President Clinton. In 1998, he received the Clarence Leroy Meisinger Award from the American Meteorological Society and, in 1999, 2000, and 2002, the NOAA Office of Oceanic and Atmospheric Research Outstanding Scientific Paper Award.

Besides being on many panels and committees and the author of close to 100 publications, about 40 of them formal, he has been associate editor of *The Monthly Weather Review*, (1994–98) and *Weather and Forecasting* (1994–97), as well as the editor of *Weather and Forecasting* (1999–present).

As the head of the models and assimilation team of the forecast research and development division at NSSL, Stensrud is responsible for coordinating NSSL's activities as they relate to improving forecasts of severe and hazardous weather events. His current research interests include the modeling and forecasting of hazardous weather, using mesoscale numerical weather prediction models, with a particular emphasis on events that occur in rather benign environments.

⊠   **Stevenson, Thomas**
(1818–1887)
Scottish
Engineer

Thomas Stevenson was born at Edinburgh, Scotland, on July 22, 1818, grandson of Thomas Smith, first engineer to Scotland's Board of Northern Lighthouses, and son of Robert Stevenson, the builder of many of Scotland's lighthouses and Margaret Isabella Balfour. Robert Stevenson is famous for creating an engineering dynasty known as the Lighthouse Stevensons. Aside from the 19 lighthouses that Robert designed for Scotland, 97 others were built between 1790 and 1940 by eight members of his family, including his son Thomas, who built 28 lighthouses with his brother David between 1854 and 1880, and three more with his nephew David between 1885 and 1886. Thomas was known as the Nestor of lighthouse illumination in Germany. He also designed harbors.

Thomas Stevenson, also an engineer to the Board of Northern Lighthouses, was particularly interested in the issues of lighting methods for lighthouses. He invented a new and improved illumination system for lighthouses, called the azimuthal condensing system of lighthouse illumination. This was a system of parabolic mirrors to concentrate the beam from a candle. Early lamps used in lighthouses produced soot, not to mention the difficulty in keeping them burning in cold and drafty areas. However, in 1822, French engineer Augustin-Jean Fresnel introduced a new lens, bullet-shaped, that could collect up to 90 percent of the lamp's light and focus it into an intense horizontal beam. Fresnel lenses are used today in most lighthouses. Thomas and Alan Stevenson, a nephew, knew of Fresnel's developments, and, in fact, Alan worked with Augustin Fresnel's brother Leonor. Thomas introduced the revolving light.

However, Thomas Stevenson is more famous in meteorology as the inventor of a thermometer screen or housing, known by his name—the Stevenson Screen. This has become the standard housing for meteorological thermometers and other instruments; it consists of a wooden cupboard or box, with a hinged door, mounted on a steel or timber stand, with the base rising about 3.5 half feet above the ground. The whole unit painted white. Indirect ventilation is provided through the bottom, double roof, and louvered sides. The instruments are placed within, and the design gives close approximation to the true air conditions but undisturbed by the effects of direct sun or terrestrial radiation.

Stevenson also contributed to refinements in rain gauges that, at the time, were proving unreliable. Remarking on how rain gauges of the time were not very accurate due to "splashing" and other designs, in 1842, he showed two forms of gauges with their mouths sunk level with the earth and with a brush around the mouth that prevented splashing from entering the gauge. Although these did not catch on because of their expense and their unworkable design in some areas, he contributed much to the discussion, and in the British Isles, a standard height for placement was established.

Stevenson married Margaret Isabella Balfour, the daughter of the Reverend Dr. Lewis Balfour, parish minister at Colinton, on August 8, 1848, at Colinton, Scotland. They had one son, Robert Louis Stevenson, born on November 13, 1850, in Edinburgh, Scotland. Although Robert started school as an engineer, he changed gears and became a writer instead, penning such classics as *Treasure Island* and *Kidnapped*. He dedicated his collections *Familiar Studies of Men and Books* to his father ". . . by whose devices the great sea lights in every quarter of the world now shine more brightly."

Along with his father, Thomas Stevenson invented the wave dynamometer, a device to measure the force of waves striking a solid surface. This was a 6-inch-diameter plate mounted on a stiff horizontal spring where a rod behind the spring would move each time a wave struck the plate.

Perhaps no one could better describe Thomas Stevenson's contributions to weather and climate than his famous son Robert. He writes a sketch about his father in his 1887 *Memories and Portraits*:

> Thus it was as a harbour engineer that he became interested in the propagation and reduction of waves; a difficult subject in regard to which he has left behind him much suggestive matter and some valuable approximate results. Storms were his sworn adversaries, and it was through the study of storms that he approached that of meteorology at large. Many who knew him not otherwise, knew—perhaps have in their gardens— his louvre-boarded screen for instruments. But the great achievement of his life was, of course, in optics as applied to lighthouse illumination. Fresnel had done much; Fresnel had settled the fixed light apparatus on a principle that still seems unimprovable; and when Thomas Stevenson stepped in and brought to a comparable perfection the revolving light, a not unnatural jealousy and much painful controversy rose in France. It had its hour; and, as I have told already, even in France it has blown by. Had it

not, it would have mattered the less, since all through his life my father continued to justify his claim by fresh advances. New apparatus for lights in new situations was continually being designed with the same unwearied search after perfection, the same nice ingenuity of means; and though the holophotal revolving light perhaps still remains his most elegant contrivance, it is difficult to give it the palm over the much later condensing system, with its thousand possible modifications. The number and the value of these improvements entitle their author to the name of one of mankind's benefactors. In all parts of the world a safer landfall awaits the mariner. Two things must be said: and, first, that Thomas Stevenson was no mathematician. Natural shrewdness, a sentiment of optical laws, and a great intensity of consideration led him to just conclusions; but to calculate the necessary formulae for the instruments he had conceived was often beyond him, and he must fall back on the help of others, notably on that of his cousin and lifelong intimate friend, EMERITUS Professor Swan, of St. Andrews, and his later friend, Professor P. G. Tait. It is a curious enough circumstance, and a great encouragement to others, that a man so ill equipped should have succeeded in one of the most abstract and arduous walks of applied science. The second remark is one that applies to the whole family, and only particularly to Thomas Stevenson from the great number and importance of his inventions: holding as the Stevensons did a Government appointment they regarded their original work as something due already to the nation, and none of them has ever taken out a patent. It is another cause of the

comparative obscurity of the name: for a patent not only brings in money, it infallibly spreads reputation; and my father's instruments enter anonymously into a hundred light-rooms, and are passed anonymously over in a hundred reports, where the least considerable patent would stand out and tell its author's story.

Thomas died on May 8, 1887, at home at 17 Heriot Row, Edinburgh.

Stolzenburg, Maribeth
(1967–   )
American
Meteorologist

Maribeth Stolzenburg was born in Albany, New York, on April 12, 1967, to Carl and Faye Stolzenburg, dairy farmers. While growing up on a dairy farm, Stolzenburg helped in farm chores, rode horses, ice skated, played piano and percussion in school and community bands, and sang in the school choir. She attended Schoharie Central School (K–12) and during high school worked as a prep-cook, waitress, and products promoter for vegetable farms. During this time, she was also a Schoharie County Spelling Bee champion and a Schoharie County Dairy Princess. She attended the State University of New York at Albany, where she received her bachelor of science degree (summa cum laude) in atmospheric science and geography in 1989. She completed her graduate work at the University of Oklahoma and received a master's degree in 1993 and a Ph.D. in 1996, both in meteorology. In 1996, she married Thomas Carlton Marshall, a professor and the chair of the physics and astronomy department at the University of Mississippi. They both work in the same field and are close colleagues as well.

While in graduate school in 1990, she published her first paper, "Characteristics of the bipolar pattern of lightning locations observed in 1988 thunderstorms," in the *Bulletin of the American Meteorological Society*. This was the outcome of her senior research project and was the winning contribution to the AMS Father James B. MacElwane Annual Award competition in 1989. The paper investigated the distinct regionalization of positive and negative polarity cloud-to-ground lightning.

Stolzenburg has made several contributions in the field of lightning research. She has found that summertime thunderstorms in the United States often display patterns of high ground-flash density of positive polarity lightning and that these thunderstorms are sometimes more intense hail producers than other thunderstorms. This research has paved the way for many detailed subsequent studies of this phenomenon.

She also has found that the electrical charge in the stratiform cloud region of mesoscale convective systems is found in horizontally extensive layers. This shows that dynamical and kinematical relationships within the MCS (mesoscale convective systems) stratiform cloud can play important roles in organizing the electrical structure. (Mesoscale convective systems are large thunderstorm complexes with convective and stratiform clouds that cover areas of 10,000 square kilometers or more.) Additionally, she has found that the height of the main negative-charge region in convective updrafts is proportional to the strength of the updraft. This research showed the importance of convective dynamics and vertical advection in electrical structure.

Stolzenburg's fourth discovery is that, independent of convection type, thunderstorm convective regions typically have one of two basic charge structures: within the updraft there are four regions of charge, and outside the updraft where there are at least six regions of charge. This permitted all in situ measurements of convective electrical structure to be organized into a comprehensive scheme. It should also allow

better understanding of the charging mechanisms and cloud parameters that are important in explaining how charge develops in thunderstorm convection.

Finally, she has determined that the two basic charge structures of convective regions can occur at the same time in a thunderstorm. This discovery shows that one basic charge structure does not necessarily evolve from the other but rather that the two may be caused by different or additional mechanisms occurring in part of the cloud.

From 1996 to 2000, Stolzenburg conducted postdoctoral work as a research associate at the University of Mississippi, where she now serves as a research assistant professor and continues her research on atmospheric electricity, thunderstorms, lightning, and mesoscale meteorology. She also has written more than 20 scientific papers and has received numerous awards for her work, including the Presidential Award for Women's Leadership in a Non-Traditional Career (1989); the Father James B. MacElwane Annual Award, American Meteorological Society (1990); and the Patricia Roberts Harris Fellowship, U.S. Department of Education (1990–93). In 2000, she received a major 18-month grant from the National Science Foundation called POWRE: Using Total Lightning Data for Storm-scale Research. She is a member of the American Meteorological Society (1989) and American Geophysical Union (1989).

Overall, Stolzenburg's scientific work has helped bring the fields of atmospheric electricity and dynamical meteorology closer together. Important ties can be made between the electrical nature (for example, charge, electric field, and lightning) and the dynamical structure of a cloud. These ties need to be documented with observations, and the observations need to be organized in an objective manner so that electricity can, in turn, be utilized better in developing a better understanding and forecast of thunderstorms.

Stommel, Henry Melson
(1920–1992)
American
Oceanographer, Physicist

Henry Stommel was born in Wilmington, Delaware, on September 27, 1920, to Walter Stommel, a chemist, and Marian Melson. Following World War I, the family moved to Sweden. His mother came back to the United States with Henry and his sister Ann, after leaving her husband and he was brought up in Brooklyn in a single-family household that contained his maternal great grandmother, grandfather, and grandmother, his aunt, and her daughter. He developed a close relationship with his maternal grandfather, Levin Franklin Melson, who shared an interest in science. He attended the public schools of New York City, including one year at the Townsend Harris High School, but finished high school at Freeport, Long Island, where the family had moved later.

He attended Yale University on a full scholarship, graduating in 1942 with a bachelor of science degree in physics. He stayed at Yale for two years, teaching analytic geometry and astronomy in the navy's V-12 program and even spent six months at the Yale Divinity School. In 1944, he began work on acoustics and antisubmarine warfare at the Woods Hole Oceanographic Institution in Woods Hole, Massachusetts, with Maurice Ewing, but he felt it did not suit him.

In 1948, Stommel published a paper in which he showed that the Gulf Stream could be explained deductively by fluid dynamics; it marked the birth of dynamical oceanography. He was the first to show how forces caused by the Earth's rotation could explain the Gulf Stream. He became known as the world's leading physical oceanographer and the father of dynamical oceanography. In 1950, he married Elizabeth Brown, a writer, church organist, and hospital chaplain. They had three children. He remained at the Woods Hole Oceanographic Institution

(WHOI) until 1959 when he left to become a professor at Harvard University. In 1963, he moved to the department of meteorology at the Massachusetts Institute of Technology (MIT), working among the likes of Jule CHARNEY. He stayed at MIT for 16 years as professor of physical oceanography and then returned to WHOI, where he remained until his last years.

Some of Stommel's greatest contributions dealt with the general circulation of the ocean, the gross thermal structure of the ocean, and the global abyssal circulation. He worked in the area of tides, attempting to explain the Coriolis force, internal waves, the general application of electromagnetic measurements to oceanic flows, the dynamics of estuaries and the related problem of hydraulic controls, and the interaction of nonlinear eddylike phenomena (hetons). With buckets of saltwater and chopped up parsley, he showed how the salinity of ocean water could affect the movement and mixing of ocean currents.

Stommel wrote more than 100 scientific papers as well as many popular articles and several books. *The Gulf Stream*, written in 1954, is considered the first true dynamical discussion of the ocean circulation. He wrote, with his wife, *Volcano Weather: The Story of 1816, the Year Without a Summer* (1983). He also wrote a book on islands that never actually existed, *Lost Islands: The Story of Islands That Have Vanished from the Nautical Charts*, (1984). Other important books include *Introduction to the Coriolis Force* (1989; with Dennis W. Moore), *View of the Sea* (1991) and *Oceanographic Atlases* (1978; with Michele Fieux).

Along with Fritz Schott, he invented a beta-spiral method for determining absolute flow in the ocean; the method has been used frequently to make estimates of the actual oceanic flow.

Starting in the late 1960s, Stommel organized a number of successful scientific cooperative programs, such as the global-scale Geochemical Sections Program (GEOSECS), the Anglo–U.S. Mid-Ocean Dynamics Experiments (MODE) and U.S.–U.S.S.R. successor POLYMODE. In 1978, he resigned from all programs and committees, feeling that too much compromising was necessary to accomplish the goals.

He was awarded the National Medal of Science; the Craaford Prize of the Royal Swedish Academy; the Agassiz Medal of the National Academy of Sciences; the U.S. National Medal of Science; the Rosenstiel Award in Oceanographic Science from the American Association for the Advancement of Science; the Sverdrup Gold Medal from the American Meteorological Society, and the Henry Bryant Bigelow Medal from Woods Hole Oceanographic Institute. He was granted foreign membership in the Royal Society of London (1983) and was elected a member of the National Academy of Sciences (1985), the Soviet Academy of Sciences, and the Académie des Sciences de Paris. Although he never sought an advanced degree, he was awarded honorary doctorates from Yale, the University of Chicago, and the University of Göteborg in Sweden.

The Stommel Research Award given by the American Meteorological Society (AMS) is named for him and is presented to a scientist for outstanding contributions to the advancement of the understanding of the dynamics and physics of the ocean.

Henry Stommel died on January 17, 1992, at age 71 but will always be known for his contributions to the dynamics of ocean currents, especially the Gulf Stream, and for his deep insight into the physics of the oceans and associated atmospheric phenomena.

⊠ **Strutt, John William**
(Lord Rayleigh)
(1842–1919)
English
Physicist, Mathematician

John Strutt was born on November 12, 1842, in Langford Grove, Essex, England, to John James Strutt, second baron Rayleigh, and Clara Elizabeth La Touche. His frail health as a boy prevented him from finishing school at Eton and Harrow, but private tutoring and four years at the Reverend G. T. Warner's boarding school at Torquay were enough to get him into Trinity College, Cambridge University, in 1861. Edward Routh, Cambridge's most famous coach and mathematician, gave Strutt a good understanding of mathematics; lectures and experiments carried out by George G. Stokes, the Lucasian Professor of Mathematics at Cambridge impressed the young Strutt.

In 1864, Strutt was awarded an astronomy scholarship, and in 1865, in the Mathematical Tripos, he was senior wrangler and also the first Smith's prizeman. (A Tripos is an examination for the B.A. degree with honors at Cambridge. The degree holder sits and disputes humorously with candidates for that degree. A wrangler is a grade of pass that is equivalent to first-class honors, with senior wrangler being the top.) In 1866, he was elected a Fellow of Trinity College, Cambridge.

After a tour of the United States in 1866, Strutt set up a lab at his 7000-acre estate at Terling Place, in Witham, Essex, and conducted experiments on the galvanometer, an instrument used to detect, measure, and determine the direction of small electric currents by means of mechanical effects produced by a coil in a magnetic field. Three years later, he published his theory of light scattering, which was the first correct explanation of why the sky is blue. It was also the year he married Evelyn Balfour (the sister of Arthur James Balfour, who later became prime minister of Britain). They had three sons. Strutt had an attack of rheumatic fever and for his health took a trip down the Nile from 1872 to 1873. He was elected as a Fellow of the Royal Society in 1873 and, during the trip, began to write *The Theory of Sound*, a book dealing with the mechanics of a vibrating medium and acoustic-wave propagation; two volumes were published in 1877 and 1878. Shortly after, his father died, and Strutt succeeded his father, becoming the third Baron Rayleigh. He was known after as Lord Rayleigh.

He was president of the London Mathematical Society from 1876 to 1878 and received their De Morgan Medal in 1890. In 1879, he wrote a paper on traveling waves, which has now developed into the theory of solitons (solitary waves). From 1879 to 1884, Rayleigh was the second Cavendish professor of experimental physics at Cambridge, succeeding James Clerk MAXWELL. He quickly organized the Cambridge laboratory into a first-rate teaching and research lab. During his tenure at the lab, he worked to standardize the ohm, a unit of electrical resistance equal to that of a conductor in which a current of one ampere is produced by a potential of 1 volt across its terminals. For his breakthrough work, he received the Royal Medal from the Royal Society in 1882, their Copley Medal in 1899, and the Rumford Medal in 1914. He returned home to his estate in 1884, resigning his position at Cambridge, but the following year, he became the Royal Society's secretary. Ten years later, Rayleigh discovered the inert gas argon, an achievement that eventually earned him a Nobel Prize in physics in 1904. He became president of the Royal Society the following year, a post he held until 1908 when he became chancellor of Cambridge University. From 1887 to 1905, he was professor of natural philosophy in the Royal Institution of Great Britain, succeeding John Tyndall. He received 13 honorary degrees.

Strutt's 446 publications in math and physics covered a wide range of work including, optics, light scattering, vibrating systems, electrodynamics and electromagnetism, density of gases, viscosity, capillarity, elasticity, sound, wave theory, color vision, flow of liquids, hydrodynamics, and photography (pinhole cameras). He established standards of resistance, current, and electromotive force. Lord Rayleigh died on June 30, 1919, at Witham, Essex. For meteorologists, he is immortalized by his work on gases, optics, and the term *Rayleigh scattering* which is the scattering of light by particles smaller than the wavelength of the light. Rayleigh scattering of sunlight from particles in the atmosphere is the reason why the sky is blue. Blue light is scattered much more than red light, so in our atmosphere, blue photons are being scattered across the sky greater than photons of a longer wavelength, resulting in one seeing blue light coming from all regions of the sky.

⊠ **Suomi, Verner**
(1915–1995)
American
Meteorologist, Engineer

Verner Suomi was born December 6, 1915, in Eveleth, Minnesota, the son of John and Anna Suomi. Aspiring to be an engineer, limited finances forced him to attend a teacher's college instead at Winona State University in Winona, Minnesota. He received a bachelor's degree in mathematics and science in 1938. Suomi taught high school science for several years, and at the outbreak of Word War II, he enrolled in a Civil Air Patrol course, where he was introduced to meteorology.

He continued his education at the University of Chicago, where he not only studied meteorology but also trained air cadets in basic meteorological forecasting. With the creation of a new department of meteorology at the University of Wisconsin in Madison in 1948, Suomi found himself as one of its first faculty members. He would spend most of his life there. He taught in the departments of meteorology and soil science and the Institute for Environmental Studies until he retired in 1986.

Suomi's doctoral thesis from the University of Chicago in 1953 dealt with measuring the heat budget of a cornfield. However, measuring the difference between the amount of energy absorbed and the amount lost in a cornfield led him to think about the heat budget for the entire planet. You needed satellites to measure this, however, and in the 1950s, satellites were becoming in vogue. On October 13, 1959, NASA launched *Explorer 7*, which carried a radiometer designed by Suomi and his colleagues. This was the first successful meteorological instrument in orbit and measured solar and terrestrial radiative energy to estimate radiation balance. The cornfield studies did not seem so unusual after all. Later, net flux radiometersondes on weather balloons showed the importance of clouds in their role in absorbing radiated solar energy.

He left Wisconsin twice to serve as the National Science Foundation's associate program director for their atmospheric sciences division in 1962; in 1964, he went to the U.S. Weather Bureau and served as their chief scientist for one year. Suomi and colleague Robert Parent, a professor in electrical engineering, went back to UW–Madison in 1965 and started the Space Science and Engineering Center (SSEC), funded by NASA and the National Science Foundation. It was here that Suomi would become a leader in satellite meteorology with several inventions. The SSEC became internationally famous for studies of the Earth's atmosphere, the planets, satellite hardware and other space flight instrument construction, and computing tools for meteorologists and space scientists.

At the SSEC, Suomi invented the "spin-scan" camera, a device that allowed weather satellites stationed in geostationary orbits to have the ability to image the Earth continuously, taking a photograph of a narrow band of the planet. He saw this placed in operation on NASA's ATS-1 in 1966, the same year he was elected to the National Academy of Engineering. This invention allowed scientists to observe weather systems as they developed, instead of using inadequate and incomplete snapshots. The camera was used on all geostationary satellites launched between 1966 and 1994. By 1967, the pictures were in color. It helped improve the accuracy of forecasting and is credited with saving many lives through the years. By 1971, he modified the Spin Scan into the Visible-Infrared Spin-Scan Radiometric Atmospheric Sounder (VAS) to observe temperature and moisture from satellites. He hoped this device would improve severe weather predictions. It was not launched until 1980 on the Geostationary Operational Environmental Satellite-4 (GOES-4), but it did perform as he predicted.

Today, Suomi's sounder technology is the only instrument that is able to observe severe storms over distances of hundreds of thousands of square miles. The original spin-scan design is no longer used in the United States, but his concept was adopted for other satellites and space probes used by NASA, National Oceanic and Atmospheric Administration (NOAA), the European Space Agency (ESA), and the Japanese Meteorological Agency.

In 1972, he introduced the Man–computer Interactive Data Access System (McIDAS), a computerized system that could take weather pictures and slow them, replay, and perform analysis. This tool was particularly useful in analyzing wind data collected during the first GARP Global Experiment (FGGE) in 1978. It is currently being used around the world.

In 1977, he received the National Medal of Science, awarded by the president of the United States. Suomi went on to use his expertise in the exploration of such planets as Venus (*Pioneer*, 1978), using net flux radiometers and other instruments. In 1980, he received the Charles Franklin Brooks Award from the American Meteorological Society. In 1983, he was awarded an honorary doctor of science degree from the State University of New York and, the following year, the Franklin Medal. In 1986, Verner was honored with a Distinguished Alumnus award from the Winona State University Alumni Society. Two years later, he was awarded the Nevada Medal by the governor of that state for his spin-scan camera invention. He continued to receive accolades until his death.

He was married to Paula (Meyer) Suomi, and they had three children. After a long bout with heart disease, he died on Sunday afternoon, July 30, 1995, at University Hospital at the age of 79. He was buried near the family home in southern Minnesota.

He considered the lack of study of water vapor to be one of the most important problems in atmospheric science. A project called SuomiNet, in his honor, is in the early stages of operation: it will link up on the Internet 100 GPS stations in universities to provide real-time water vapor data online. Scientists hope that by having improved water-vapor information, our understanding of floods, severe storms, and regional and global climate change will be improved. The project is funded by the National Science Foundation. Verner Suomi is internationally recognized today as the "Father of Weather Satellites."

Antony J. Surtees. He continues his research activities in the field of atmospheric electricity, which includes lightning and surge protection—a subfield of the broader field of meteorology. *(Courtesy of Antony J. Surtees)*

⋈ **Surtees, Antony J.**
(1959–)
Australian
Physicist, Engineer

Antony J. Surtees was born on February 14, 1959, in Salisbury, Rhodesia (now Harare, Zimbabwe). His great grandfather was a member of the British Pioneer column that settled the country in 1893 under Cecil John Rhodes. This British imperialist and business magnate is perhaps best remembered as the benefactor of the Rhodes Scholarship to Oxford University and the founder of the De Beers Company, the world's largest diamond producer.

Surtees's early childhood was spent on a farm where his parents were tobacco farmers—tobacco being one of the most important national exports of Zimbabwe. Soon after his two brothers were born, the family moved nearer to the city so that he could attend school. He was educated under the British O-, M-, and A-level Cambridge school system. His early interest in electronics and

anything mechanical stemmed from his father's love of fixing anything and everything—a skill finely honed with the regular bouts of temperamental old farm tractors! He can recall with joy the many happy years spent fiddling in his father's workshop with his faithful dog by his side. He attended Alan Wilson Technical High School, where he was awarded the engineering and electronics prizes for his A-level year. This confirmed his desire to become an engineer on leaving school, and in 1979, he was accepted to study electrical and electronic engineering at the University of Cape Town (UCT) in South Africa.

Surtees was awarded the Argus Scholarship in 1980 and on graduating in 1982 went to work for Argus, his sponsoring company. In 1984, he left his position of assistant chief engineer to return to study toward his Ph.D. under Professor John F. W. Bell, renowned for his research in the early days of radar during World War II and later for his work in transducers. A transducer is a substance or a device, such as a piezoelectric crystal, a microphone, or a photoelectric cell, that converts input energy of one form into output energy of another. In 1987, on receiving his Ph.D. in vibrating transducer design, Surtees accepted a position with the British Plessey Electronics company at its Cape Town facility. Here his work included research into a differential GPS (gobal positioning satellite) system for hydrographic dredging (waterways for example), a mine hoist monitor, and a new electronic detonator for use in the South African mining industry. For his last project, he developed an application-specific integrated circuit (ASIC).

At the end of 1990, Surtees and his family immigrated to Australia where he joined Global Lightning Technologies (GLT) in Hobart, Tasmania. GLT was considered one of the leading companies in the manufacture of lightning-protection air terminals and surge-protection devices. In 1991, he participated in the company's rocket-triggered lightning experiments at NASA's Cape Kennedy LP research facility. This field research program involved the instrumentation of an array of different air terminals (lightning-rod devices), including conventional Franklin rods, active early streamer emission devices, and delayed triggering (corona-inducing, charge-dissipation) devices to monitor their streamer initiation properties under the dynamic conditions of rapid E-field (electric-field) changes.

He has spent much of his time working in Asia where the very high isokeraunic (number of days of lightning activity each year) levels of equatorial lightning are an ideal test bed for evaluating the company's products.

Following the acquisition of GLT in 1997 by ERICO Inc., Surtees accepted a transfer to the parent company's head office in the United States. He now holds the post of director of standards and technology and is worldwide product manager for the surge-protection product line. He is a recognized speaker on the subject of lightning protection and is frequently requested to conduct technical seminars around the world. He is a U.S. delegate to various standards-setting bodies and working groups including IEC/SC37A (low-voltage surge-protective devices), IEC/TC81 (lightning protection), NEMA 5VS (low-voltage surge-protective devices), IEEE SPDC, US ICE Technical Advisor Group (TAG), and the NFPA (National Fire Protection Association). IEC is the International Electrotechnical Commission. NEMA is the National Electrical Manufacturers Association. He is a Fellow of the Institution of Electrical Engineers U.K.; a senior member of the Institute of Engineers, Australia; a member of the Institute of Electrical and Electronic Engineers, United States; and a chartered professional engineer. He continues his research activities in the field of atmospheric electricity, which includes lightning and surge protection—a subfield of the broader field of meteorology.

⊠   **Sverdrup, Harald Ulrik**
(1888–1957)
Norwegian
Oceanographer, Geophysicist

Harald Ulrik Sverdrup was born in Sogndal, Norway, on November 15, 1888, to Johan Edvard Sverdrup and Maria Vollan. His education included a military academy in Oslo in 1907–08 and, shortly after, a stint at the University of Oslo, studying astronomy and geography. In 1911, he became an assistant to the famed meteorologist Vilhelm BJERKNES. He joined other students such as Carl-Gustaf ROSSBY and Theodore Hesselberg, a group that later became known as the Bergen School of Meteorology and a major influence on the newly emerging field.

His teacher Bjerknes moved on to Leipzig in 1913 to become professor of geophysics at the University of Leipzig and director of the newly founded Leipzig Geophysical Institute. Assistant

Sverdrup followed and completed his dissertation there in 1917, on North Atlantic trade winds, as well as writing and publishing some 20 papers on atmospheric and oceanic physics. At the outbreak of World War I, as Norwegian scientists moved back to Norway, Sverdrup accepted a post as chief scientist on Roald Amundsen's 1918 Arctic expedition on the sailing vessel *Maud*. (Amundsen already made it to the South Pole in 1913.) The 1918 *Maud* trip was set out to follow a plan of polar drift in the Arctic. The three-year study ended up lasting seven and one-half years. Sverdrup studied theories of tidal currents and dynamics, atmospheric electricity, magnetics, and physical oceanography on the Siberian shelf. Up to that the time, the *Maud* expedition was the largest undertaking to a polar region, with its huge assortment of equipment for measuring oceanographic and meteorological data. But the project failed. The ship froze in the Arctic coastal ice and was stranded for two winters. After repairs, in June 1922, it again froze near Wrangel Island northeast of the former USSR. It ended up on the continental shelf off northeast Siberia and remained there for three years.

Although it did not meet their original goals, the scientific work of Sverdrup earned the expedition the reputation of being one of the most important Arctic research projects ever carried out. Sverdrup also acted as an anthropologist during his years in the Arctic, spending eight months with the Siberian Chukchi, or Russian *Chauchu*, which means "rich in reindeer." The Chukchi are the largest Native nation (currently about 15,000) on the Asian side of the North Pacific. He wrote about his experiences in a book in 1938, *Hos Tundra-Folket*.

Sverdrup wrote up his expedition observations into scientific reports and contributed more than two-thirds of the Maud expedition report. His writings increased his reputation as an Arctic and oceanographic scientist, although he admitted that he feared his work might not have been up to standards because he was out of contact with fellow scientists during the long voyage on the *Maud*. He continued research on conditions in the sea and the atmosphere, including tides, currents, circulation in the Arctic and the Pacific, and the trade winds.

While awaiting repairs to the propeller of the *Maud* during the winter of 1921–22, he worked at the Carnegie Institute in Washington, D.C., on magnetism and Arctic tidal dynamics. On his return from the *Maud* trip, he succeeded Bjerknes as professor of geophysics at Bergen University in 1926. Bjerknes had joined the faculty of the University of Oslo and, while there, Sverdrup married Gudrun Bronn Jaumund on June 8, 1928.

In 1931, Sverdrup once again left Bergen to take a research position at the Christian Michelsens Institute in Bergen to follow up on the research of the Maud expedition. During this time, he attempted to cruise beneath the North Pole in a World War II surplus submarine, purchased for a dollar by Hubert Wilkins and named *Nautilus*. The sub broke down, and the expedition failed. It did not stop him from continuing his research interests. He spent two months sledding across Spitsbergen in 1934 with H. W. Ahlmann, studying boundary layer processes.

In 1936, he moved to California to become director of the Scripps Institution of Oceanography. Under his leadership, Scripps became a modern oceanographic institution with a high reputation. Scripps conducted the first comprehensive hydrographic survey of the Gulf of California.

In 1942, Sverdrup and coauthors Richard H. Fleming and Martin W. Johnson wrote the first modern textbook in oceanography entitled *The Oceans: Their Physics, Chemistry and General Biology*. It was considered so accurate that the publication was delayed during the war, and when published, it had export restrictions. That year, he also authored the 246-page book *Oceanography for Meteorologists*.

Several of the famous Bergen School scientists were in America during the time of World

War II at MIT and UCLA while Sverdrup was at Scripps. Sverdrup, Rossby, Bjerknes, and Jorgen Holmboe established a training school for military meteorologists at UCLA. UCLA Department of Meteorology and Scripps, under Sverdrup, trained more than 1,200 military weather officers between 1941 and 1945. Many of the graduates conducted weather forecasts and surf predictions for allied landings in Normandy, North Africa, and the South Pacific, saving countless lives. This important application of meteorological knowledge is used even today in military planning.

He returned to Oslo in 1948 as director of the Norwegian Polar Institute, and in 1949, he became professor of geophysics at the University of Oslo, later becoming the school's vice-chancellor. He went back to the Arctic, reformed the Norwegian curriculum, and chaired the Norwegian relief program in India in 1952.

During his lifetime, Sverdrup was a member of numerous scientific organizations such as U.S. National Academy of Sciences and the Norwegian Academy of Sciences. He also was president of the International Association of Physical Oceanography (1948–51) and of the International Council for the Exploration of the Sea and was vice president of the American Geophysical Union (and chairman of its Division of Oceanography, 1944–47).

Recognized for his work by many organizations included his receiving the Agassiz Medal from the National Academy of Sciences in 1938, the Patron's Medal of the Royal Geographical Society "for contributions to polar exploration and for oceanographic investigations" in 1950, and the American Geophysical Union William Bowie Medal in 1951. He became a member of the Swedish Order of the North Star (*Kungliga Nordstjärnesorden*). The order, instituted February 23, 1748, by King Frederik I, is awarded to foreign and Swedish citizens for meritorious achievements in science and literature, as well as for service to the state and society.

Sverdrup died suddenly in Oslo on August 21, 1957. In his honor, the American Meteorological Society awards The Sverdrup Gold Medal to researchers who make outstanding contributions to the scientific knowledge of interactions between the oceans and the atmosphere. The sverdrup (1 sverdrup (Sv) = 106 m^3 s-1) is a unit of volume transport named for him. The Sverdrup Islands in Arctic Canada are named after him.

⊠ Swanson, Brian Douglas
(1951–   )
American
Physicist

Brian Swanson was born on February 27, 1951, in Seattle, Washington, to Allan Swanson, a dairy farmer, and Florice Swanson, an accountant. Swanson attended Bothell High School with interests in sports, music, philosophical questions, the desire to build things, and the workings of machinery. Beginning in 1969, he attended the University of Washington at Seattle, receiving his B.A. in philosophy in 1973. After working as a Peace Corps volunteer with the Rural Electrification Administration in the Philippines, he returned to the University of Washington for a bachelor of science degree in 1981 in physics and another bachelor of arts degree in mathematics the same year. He received a master's degree in physics in 1986, the same year he married Mary Laucks, a fellow physics graduate student. He continued at the University of Washington and received a Ph.D. in 1992, also in physics.

Swanson's major interests are phase transition physics, atmospheric and cloud physics, ice-particle microphysics, and ice physics. He published his first paper in 1989, "Synchrotron Studies of the First-Order Melting Transitions of Hexatic Monolayers and Multilayers in Freely Suspended Liquid-Crystal Films." The focus of this work was to explore how this phase transition occurs and whether the melting is first-order or continuous—he found in this case a hysteretic first-order transition. Swanson's research has led to discoveries that surface-induced order is important in some phase transitions of liquid crystals. In his cloud physics experiments, he has shown that ice-particle breakup during sublimation may be important to secondary ice production in clouds. He also made some of the first light-scattering measurements from well-oriented ice particles and has found that ice-particle-initiation mechanism (whether a particle starts as a frozen droplet or a small ice seed) is important in determining ice-particle shape. This work is important to our understanding of ice-particle concentrations in clouds and in trying to quantify the light-scattering properties of the most common shapes of atmospheric ice particles.

Swanson currently is a research assistant professor in earth and space sciences at the University of Washington at Seattle, where he and his colleagues continue research on the study of aerosol properties, aerosol-cloud interactions, cloud electrification and lightning characteristics, ice physics, planetary atmospheres, and the role of snow and clouds in climate. He has published some 18 articles in phase transition physics, atmospheric and cloud physics, ice-particle microphysics, and ice physics.

Swanson and his students are currently working in five areas. They have been studying the growth, sublimation, and scattering properties of tropospheric ice particles. One of the instruments they use is a temperature- and humidity-controlled electrodynamic balance (EDB) that allows them to observe single ice particles away from the influence of surfaces. Their work focuses on growing and sublimating both habited and unhabited ice particles in the laboratory under controlled conditions similar to those observed in nature.

Second, they have been developing a droplet freezing instrument to observe the freezing of supercooled droplets in free-fall. This instrument makes possible the observation of the freezing process away from the influence and possible contamination of substrates. One study looked at the microphysics of homogeneous and heterogeneous freezing of bionucleants, the freezing temperatures, and the associated droplet-shape changes induced by the freezing process. They are also looking at the influence of inorganic, organic acid solutions and biogenic ice nuclei on the process of heterogeneous ice nucleation.

Swanson's group is also interested in the microphysics of charge transfer at ice surfaces—the process responsible for thunderstorm electrification. They have built an instrument to observe the fluorescence from a pH-sensitive molecular probe to observe the pH profile near a growing or sublimating ice surface. (pH is a measure of the acidity or alkalinity of a solution, numerically equal to 7 for neutral solutions, increasing with increasing alkalinity and decreasing with increasing acidity. The pH scale commonly in use ranges from 0 to 14.)

Another interest is the microphysics of uptake for various atmospheric trace impurities into ice particles. They are developing resonant Raman spectroscopy techniques and EDB instrumentation to study the uptake of acid gases and other impurities into ice of various shapes and defect densities.

Finally, Swanson's group has begun a side project that is looking at ice-crystal formation on plant surfaces. During a trip to northern India, Swanson observed remarkable sap crystals forming on the surface of a plant of the Dock family. He has recently found other northwestern U.S. trees that exhibit sap crystal formations with a distinct and different morphology and is studying the formation mechanism for these crystals. He continues making contributions in the study of aerosol properties, aerosol-cloud interactions, cloud electrification and lightning characteristics, ice physics, planetary atmospheres, and the role of snow and clouds in climate.

⊠ **Symons, George James**
(1838–1900)
English
Climatologist

George James Symons was born on August 6, 1838, in Queen's Row, Pimlico, London, an area near Chelsea and Westminster. He was a meteorologist before he was out of his teenage years. He joined the Royal Meteorological Society in 1856 at 17 years of age, and the following year, became the meteorological reporter to the Registrar General, a job he held his entire life. Shortly after, he created a network of voluntary observers in 1860 to study Britain's rainfall; the network evolved into the British Rainfall Organization. By 1899, it had more than 3,000 volunteer observers. Under his 40-year leadership that promoted accurate and reliable collection methods, the organization collected volumes of rainfall data, and he authored a series of annual volumes on the rainfall of the British Isles during this time. In 1919, the organization was absorbed into Britain's Meteorological Office (known as the Met Office).

In 1863, he founded and edited a monthly publication about rain that three years later became *Symons' Monthly Meteorological Magazine*. Like the British Rainfall Organization, this magazine became part of the Met Office and evolved into *The Meteorological Magazine*, the Met Office's official publication which was published until December 1993.

Symons, who believed in obtaining accurate meteorological data, invented or improved several instruments, including the brontometer, a device for recording the sequence of phenomena in thunderstorms, and a storm-rain gauge for measuring heavy but short rainfalls. He is also given credit for promoting the Stevenson screen to be accepted as the standard housing for meteorological instruments. A Stevenson Screen is a side-louvered boxed shelter that contains temperature and relative-humidity equipment that shields them from sunshine and the elements but allows free movement of air. The screen design, now the world's standard, was invented by Thomas STEVENSON, a civil engineer who is also the father of author Robert Louis Stevenson.

During his lifetime, Symons became more of a collector than a writer; although he wrote hundreds of articles, he compiled a collection of thousands of meteorological books that he left to the Royal Meteorological Society. On June 6, 1878, he was elected a member of the Royal Society and was a member of their Krakatoa Committee. Ten years later, he would publish a book called *The Eruption of Krakatoa and Subsequent Phenomena*, about the famous August 26, 1883, eruption of the island volcano of Krakatau ("Krakatoa") in Indonesia. In 1880 and again in 1900, he became president of the Royal Meteorological Society but died before completing his second term. He was also a member of the Scottish Meteorological Society.

After his death in 1900, the Royal Meteorological Society created a memorial fund and biennial gold medal in his honor, which today is the society's highest award. It is given to those who distinguish themselves in connection with meteorological science. The medal's logo is a representation of the Tower of the Winds at Athens, also known as the Horologium of Andronikos Kyrrhestes, built in 40 B.C. by the astronomer Andronikos. It was adopted as the society's logo in 1902. Previous winners of the Symons Medal include notable meteorological figures such as Cleveland ABBE and Sydney CHAPMAN. Symons died on March 10, 1900, and was buried in Kensal Green cemetery, in London.

T

Teisserenc de Bort, Léon Philippe
(1855–1913)
French
Meteorologist

Léon Philippe Teisserenc de Bort was born in Paris on November 5, 1855. He established his own observatory at Trappes (1896) and was a pioneer in the use of unmanned weather balloons, fitted with meteorological equipment to investigate the atmosphere.

The first scientific balloon flight was conducted in 1783 by Jacques Charles, a French physicist, and ascended in a hydrogen balloon, carrying a barometer and thermometer. The use of sounding balloons for meteorological observation was first introduced in 1754. In March 1892, French aeronaut Gustave Hermite was the first to launch *ballons perdus* (lost balloons).

However, by 1902, using homemade sounding-balloons filled with hydrogen gas, and using his own instrumentation, including barometers and thermometers to measure temperature and pressure, Bort had made more than 500 ascents, including more than 200 flights that went above 7 miles (11 km). His observations revealed that the atmosphere could be divided into succinct layered units based on temperature profiles, and he named this upper part of the atmosphere the *stratosphere* (Greek for "sphere of layers"), reasoning that at this height, different gases would exist in distinct layers and not mingle. He named the lower part of the atmosphere the *troposphere* (Greek for "sphere of change") because it was here that there were constant temperature changes and mingling of atmospheric gases.

To separate these two layers of the atmosphere, Bort named the division between them as the *tropopause*, and his balloon experiments indicated that this divisional layer was 7 miles (11 km) high over France.

He presented his research on April 28, 1902, to the Paris Academy and the following month to the Berlin conference of the International Aerological Commission. Bort died in Cannes on January 2, 1913.

Thomson, William
(Lord Kelvin)
(1824–1907)
Irish
Mathematician, Physicist

William Thomson was born in College Square East, Belfast, Ireland, on June 26, 1824, to James Thomson, a mathematician and professor of engineering in Belfast, who brought William and six other children up alone, as his wife died when William was six. When William was eight, his father was appointed chairman of the math department at the University of Glasgow, and young William became knowledgeable in math. He would grow up to become one of the greatest minds of his time and would help establish modern physics.

He attended Glasgow University at the age of 10 and studied astronomy, chemistry, and physics, but it was obvious that his father's math background would rub off on young William and last for most of his life. At 15, he wrote an *Essay on the Figure of the Earth* that won him a gold medal from the university. Two years later, in 1841, he entered school at Cambridge and published his first paper, "Fourier's expansions of functions in trigonometrical series." The following year, "On the uniform motion of heat and its connection with the mathematical theory of electricity" was published. Having read Jean-Baptiste Fourier, Thomson was quick to defend him when Fourier came under attack for his theories.

After college in 1845, Thomson went to Paris and found work in the laboratory of Henri-Victor Regnault, the French physicist and chemist. His stay was short, thanks to his father who lobbied for young William when a position opened at Glasgow. In 1846, he returned to Glasgow to become a professor of natural philosophy (physics), a position that he kept for 53 years. It was here that he began a long and distinguished career and a 50-year relationship with another mathematician, George Stokes, starting in 1847. They corresponded about various studies that dealt with heat and fluids.

Thomson established the science of thermodynamics and formulated the first and second laws. These studies led him to develop a new "absolute" temperature scale in 1848, where the freezing point of water on this scale is 273.15 degrees and the boiling point as 373.7 degrees. This temperature scale was later called Kelvin because Thomson received the title Baron Kelvin of Largs from the British government in 1892. The scale was name for him.

Shortly after his absolute temperature scale was established, he was elected to the Royal Society of London in 1851 and received its Royal Medal in 1856. Even with all his scientific theories, it was an invention in the 1850s that gained him enough wealth to live as a "lord." He designed a mirror-galvanometer that was used in the first successful sustained telegraph transmissions in a transatlantic submarine cable, laid between Ireland and Newfoundland in 1858–65. He was knighted in 1866 for this work. He made enough money to own a palatial estate and a yacht.

Thomson teamed up with Scottish physicist and mathematician Peter Guthrie Tait to produce the *Treatise on Natural Philosophy*, on which they began to work in the early 1860s. Only two volumes which covered kinematics and dynamics, were ever written, but they were standard texts for many years. Thomson served as president of the British Association for the Advancement of Science in 1871.

Thomson developed a dynamical theory for heat. This was explained in his 1856 paper to the Royal Society, entitled "Dynamical Illustrations of the Magnetic and Helicoidal Rotary Effects of Transparent Bodies on Polarised Light." A theory followed for electricity and one for magnetism later, they led James Clerk MAXWELL to his theory of electromagnetism in 1873, considered to be one of the greatest achievements of 19th-century physics. Thomson was a mentor to Maxwell.

Thomson received the Royal Society's Copley Medal in 1883 and served as president from 1890 to 1895. He also belonged to the Royal Society of Edinburgh, serving three different terms as president of this society between 1873 and his death in 1907. Thomson published more than 600 papers and received patents on more than 70 inventions. Among his patents were ones for a compass (adopted by the British Admiralty); sounding equipment; and an analog computer for measuring tides in a harbor and for calculating tide tables for any hour, for any date. He established a company with Glasgow instrument-maker James White to manufacture these items and a number of electrical measuring devices. In 1881, he had the first house in Scotland with electric lighting.

He did not always find himself on the same side as his colleagues on various issues, such as the age of the Earth, Darwin's evolution, and other important topics of his time. He is remembered for making the serious faux pas in 1900 to a gathering of physicists at the British Association for the Advancement of Science. He stated that there was nothing new to be discovered in physics except for more-precise measurements. He died on December 17, 1907, at his estate at Netherhall (near Largs), Ayrshire, Scotland, and was buried in Westminster Abbey, London.

⊠   **Tierney (née Morris), Heidi Ellen**
(1970–   )
American
Physicist

Heidi Tierney was born on June 26, 1970, in Dallas, Texas, to Katherine Anne Bachman, an ice skater. As a youngster, her interests included amateur astronomy, jazz saxophone, community theater, and art as she attended High School in Mineola, Texas, and worked as a food preparer and server at a small family-owned restaurant in Mineola called Just Burgers.

She attended the University of Texas at Austin while working for the city of Austin as a security guard for the city power plants and received a B.A. in astronomy in 1993. For her graduate work, she attended the University of Oklahoma at Norman and received her master's degree in physics in 1996. She received a Ph.D. in physics in 2002 also from the University of Oklahoma.

Tierney has already made discoveries in VHF (very high frequency, 30-300 MHz) lightning observations by assigning locations on the Earth's surface to VHF-radio signals received by the FORTE satellite. FORTE was launched on August 29, 1997, and is in a low Earth orbit at about 800 kilometers. It carries instruments designed to record optical light pulses and radio waves produced by lightning, and to date, millions of signals have been recorded. If a radio signal is recorded by one of FORTE's receivers, it is not automatically known from where on the surface of the Earth the signal originated. Some of the satellite-recorded signals have locations assigned to them by using additional receivers on the Earth, such as the National Lightning Detection Network (NLDN). Tierney showed that by analyzing the ionospheric distortion of the signals, one could figure out which thunderstorm produced a given cluster of radio signals recorded by the satellite. This work resulted in her first publication in 2001, "Determination of source thunderstorms for VHF emissions observed by the FORTE satellite," with A. R. Jacobson, W. H. Beasley, and P. E. Argo in *Radio Science*.

The FORTE satellite often records pairs of pulses in the VHF radio range. The two pulses are separated by 10 to 30 microseconds. At one time, it was tested to see if the second pulses might emanate from lightning-related events known as sprites and jets; in this scenario, the first pulse originates within the thunderstorm, and the second is produced by a sprite or a jet. However, it is now believed that a single "event" is responsible for both signals. A complete physical description of what this event might be is currently under investigation. This is, in part, because the time separation of the two pulses is consistent with one observation of a radiation pulse that takes a direct path to the satellite and a second pulse of radiation that is reflected by the Earth's surface. Tierney's second paper, "Transionospheric pulse pairs (TIPPs) originating in maritime, continental and coastal thunderstorms: Pulse energy ratios," published in 2002, showed that thunderstorm VHF pulses (also observed using FORTE) that originate over water have stronger surface-reflected signals than those that occur over land.

Heidi Ellen Tierney. Tierney is currently involved in research that is exploring whether cosmic-ray secondary electrons might be involved in the initiation of lightning. *(Courtesy of Heidi Ellen Tierney)*

The inferred reflection coefficients from the data are in agreement with reflection coefficients expected for land and seawater. This work helped confirm that, for many TIPPs, the second pulse of a pair is a ground-reflected signal. The location of the signals received by FORTE (based on her first paper) made this discovery possible.

Tierney is currently involved in research that is exploring to see if cosmic-ray secondary electrons might be involved in the initiation of lightning. Her current research will aid in our understanding of VHF-radio radiation from electron avalanches that are initiated by cosmic ray secondary electrons. Her colleagues have already shown that models of "runaway" electron avalanches in thunderstorm electric fields can reproduce observations of X rays associated with lightning. Tierney is testing whether the peak

signal strengths and spectra from models of these electron avalanches can also reproduce radio observations made by ground-based receivers or by those on board the FORTE satellite. This is the subject of her doctoral work. She believes that scientists are very close to a more complete understanding of lightning because of the dedication and cooperation of contemporary observationalists and theorists. She currently is working on this problem at Los Alamos National Laboratory.

In April 2000, she married Tom Tierney, a high-energy density physicist at Los Alamos National Laboratory. They met in Los Alamos when they began to work as graduate research assistants in 1998.

⊠ Tinsley, Brian Alfred
(1937–   )
New Zealand
Physicist

Brian Tinsley was born on April 23, 1937, in Wellington, New Zealand, to Terence Anders Tinsley, a construction contractor, and Nola Emma Tinsley. He attended elementary school by correspondence because of his susceptibility to asthma and colds at that time, and his mother acting as a tutor, then he studied at Plymouth Boy's High School, with interests in reading and mountain climbing. He attended the University of Canterbury, New Zealand, and obtained a bachelor of science degree in 1958, a master's degree in 1961, and Ph.D. in 1963, all in the field of physics.

Tinsley has made major contributions in three areas. He discovered structures in the upper atmosphere and ionosphere that are identified by airglow observations. He found that low-latitude auroras are caused by energetic neutral atom precipitation, and he identified weather and climate response to solar activity, interpreted as cloud microphysical responses to changes in electrical current density in the global electric circuit. Tinsley's research has led to improved understanding of atmospheric processes and has improved capabilities for predicting climate change.

Tinsley has been actively involved in aeronomy, the observational and theoretical research on upper atmosphere processes, for more than 40 years and has served on many national and international organizations in this field.

From 1986 to 1988, while serving as program director for aeronomy at the National Science Foundation, he had the opportunity to discuss longstanding problems in atmospheric science with program directors in areas of meteorology. This led him to begin research on the centuries-old question of the effects of changes in the sun on day-to-day weather, year-to-year climate changes, and global warming on the century time scale. From 1990 to 2002, he published more than a dozen papers on his developing theory of a mechanism for such solar effects. Tinsley's theory asserts the involvement of the solar wind, in contrast with the traditional view that changes in solar brightness, are responsible. The solar wind is a highly conducting, extremely hot gas that blows from the sun outward over the Earth. It impedes the flow of high-energy cosmic-ray particles coming in from the galaxy and energizes high-energy electrons in the Earth's radiation belts that precipitate into the atmosphere; both of these effects change the column conductivity between the ionosphere and the Earth's surface. Third, the solar wind changes the potential difference between the ionosphere and the Earth in the polar cap regions. All three effects alter the ionosphere-Earth current density (Jz) that is part of the global atmospheric electric circuit and that flows down from the ionosphere to the surface and into and through clouds. There are good correlations, on the day-to-day time scale, between the three solar wind-modulated inputs described and small changes in atmospheric temperature and dynamics.

Tinsley's theory is that the atmospheric response is due to changes in the rate of ice pro-

duction at the upper surfaces of clouds, caused by the changes in current density into the clouds. Ice production has consequences both for cloud thickness and reflectivity to sunlight, precipitation rates and latent heat transfer of clouds. Both sets of characteristics are capable of affecting global atmospheric temperature and dynamics. About half of the global warming during the past century appears to be due to changes in the sun and the solar wind. Annual and decadal climate changes attributable to the Sun are significant compared with climate changes due to other

Brian Alfred Tinsley. His major contribution in meteorology is the discovery of the importance of atmospheric electrical effects on the microphysics of clouds that are only weakly electrified, with consequences for externally forced climate change. *(Courtesy of Brian Alfred Tinsley)*

sources. This mechanism also explains many reports of high rates of ice formation in certain types of clouds, which has been a long-standing puzzle for cloud physicists.

Tinsley, working with Dr. K. A. Beard at the University of Illinois, has investigated various pathways for electrical effects on ice production in clouds. Their theoretical analyses that utilize laboratory and in-situ cloud data point to the importance of such processes for production of ice where there are electrically charged, evaporating droplets in clouds. In 1998, Tinsley, working with Beard and others, made numerical models of a process called electroscavenging (electrically enhanced collection of evaporation nuclei by nearby droplets). The results show that the electrostatic charges on the residues of the evaporated droplets greatly increase the rate at which the residues (evaporation nuclei) make contact with supercooled droplets, which are then likely to freeze. This increase is due to the hitherto overlooked electrical effects of induced image charges. The group's findings overthrow a long-standing assumption by cloud physicists that electrical effects in clouds are unimportant except in thunderstorms.

Electroscavenging followed by contact ice nucleation is likely to be more important for clouds with broad-droplet-size distributions, high liquid water content, and cloud tops at temperatures just below freezing. Low-level marine stratocumulus clouds have these characteristics.

Tinsley has published more than 100 papers in his 40 years of research, funded by NSF and NASA. He first married Beatrice Muriel Hill in 1961; in 1977, he married Yvon Gwyn Reeder, and they have four children.

His major contribution in meteorology is the discovery of the importance of atmospheric electrical effects on the microphysics of clouds that are only weakly electrified, with consequences for externally forced climate change. He currently is a member of the faculty of the University of Texas at Dallas.

⊠ **Todd, Charles**
(1826–1910)
English
Astronomer, Meteorologist

Charles Todd was born July 7, 1826, in Islington, London, England, the son of George Todd, a grocer from Greenwich. A math prodigy with interests in astronomy, he became a "supernumerary computer" (human calculating machine) at age 14 at the Royal Observatory in Greenwich (1841–47) and then an Assistant Observer at the Cambridge University Observatory (1848–54). He returned to the Royal Observatory for a short time (1854–55), working in the electrical research department.

The year 1855 was a pivotal year for the 30-year-old astronomer. He married 17-year-old Alice Gillam Bell, daughter of Edward Bell of Cambridge, before he left for South Australia to become the superintendent of Telegraphs and Government Astronomer and Meteorologist. He also was placed in charge of the new Adelaide Observatory (a post he held until he retired in 1906). Astronomer Sir George Airy, director of the Cambridge Observatory where he had worked previously, recommended him for the job.

When he and his wife arrived in Australia on November 5, 1855, James McGeorge had finished placing a telegraph line between Adelaide and Port Adelaide that very day. However, Todd quickly initiated the construction of miles of telegraph lines, that by 1858, a line stretched to Victoria and New South Wales. He is most famous for the 2,000-mile *Overland Telegraph Line* from Adelaide to Darwin, consisting of 36,000 poles (4 feet deep, 15 feet tall), running south to north through the continent and finished in only two years (by 1872). John Ross, who in turn followed the route set out by surveyor John McDouall Stuart in 1862, plotted the overland route for him. This line linked Australia to an undersea cable from Indonesia to Port Darwin and made communication between Australia and the rest of the world possible.

Because he was superintendent of telegraphs, he used the new technology and developed a meteorological network in 1856 in which all his telegraph operators began to transmit meteorological observations to him at the observatory. A similar project started only a few years earlier by Joseph HENRY in America. Because Henry had corresponded with Greenwich, Todd may have learned about the project from friends at Greenwich because he worked there as recently as 1847. Todd's meteorological network reached South Australia, the Northern Territory, and Western Australia. This system made it possible for him to produce some of the earliest synoptic weather maps, later adopted elsewhere, and to produce weather bulletins.

In addition to his post as the astronomical observer and superintendent of telegraphs, he became postmaster general of South Australia in 1870 and continued that post until 1906. Although it appears that he had no formal education, he did receive an honorary Masters of Arts from Cambridge University in 1886.

Todd is also credited for another important meteorological observation. In 1888, he recognized the connection between droughts in Australia and in India due to a phenomenon known today as El Niño, or the Southern Oscillation. A Fellow of the Royal Astronomical Society since 1864, he made important astronomical observations during his years as an astronomer. Some of these include viewing the transit of Venus in 1874 and 1882 (when Venus passes in front of the Sun as seen from Earth), and Jupiter's satellites. He was one of the first observers of the planet Neptune, and he determined Australia's longitudes.

An astronomer to the end, he helped start the Astronomical Society of South Australia. The first meeting, which he chaired, was held on December 22, 1891, and he delivered the first paper, the inaugural address, in June 1892, titled "Two British Astronomers, Their Lives and Labours." The society became the Royal Society's astronomical section on May 18, 1892. Todd

remained its president until his death. The Adelaide Observatory was destroyed after World War II. Todd was elected Fellow of the Royal Society on June 6, 1889, the same year his daughter Gwendoline married Sir William Bragg, a physicist who went on to win the Nobel Prize in physics in 1914.

In 1893, he was knighted for his earlier telegraph work. He remained in charge of the Department of Posts and Telegraphs until January 1, 1901, when the six colonies of Australia came together to form a federation; he then took the title deputy postmaster general.

Alice and Charles had a total of six children. His wife Alice died in 1898, and he retired in 1906. Four years later, he died on January 29, 1910, at The Esplanade, Semaphore. He is buried next to his wife Alice at the North Road Cemetery. On his death, son-in-law Sir William Bragg wrote a tribute for the Royal Society in London that reads: "He had no commanding personality; at first glance it might have been difficult to discover the source of his power. He was clearly a bright and happy man—kind, generous, full of vitality, with a perfectly boyish sense of fun."

After his Overland Telegraph passed through the town of Stuart, the local river (now dry) was named Todd River in his honor, and the nearby springs was named Alice Springs for his wife. In 1933, Alice Springs was officially adopted as the name of the town. Today, the Charles Todd Medal for Excellence in Telecommunications is the highest individual honor awarded annually by the Australian telecommunications industry.

Torricelli, Evangelista
(1608–1647)
Italian
Physicist, Mathematician

Evangelista Torricelli, Italian physicist and mathematician, was born in the northern town of Faenza on October 15, 1608. Torricelli is credited

Evangelista Torricelli. Torricelli is best known as the inventor of the barometer. *(Courtesy of NOAA Image Library)*

with inventing the barometer and was the first to create a vacuum.

Torricelli was one of three sons of Gaspare Torricelli, a textile artisan of modest means who sent his young son to his uncle, a Camaldolese monk, for his education. He attended the Jesuit College of Faenza in 1624, and in 1627, he traveled to Rome to attend the Collegio Romano. He was a good student and found himself being taught by Antonio Benedetto Castelli, professor of mathematics at the University of Sapienza, Rome, and a good friend of Galileo GALILEI. Castelli hired him as his secretary and substitute teacher from 1626 to 1632, and from there to 1641, Torricelli appears to have worked for Mon-

signor Giovanni Ciampoli, another friend of Galileo, and perhaps other professors.

In 1641, Galileo invited Torricelli to become his literary assistant and secretary in Pian dei Giullari, Florence, after reading Torricelli's "Trattato del Moto" (Treatise on Motion), his thesis amplifying Galileo's work on projectiles. Castelli had sent it to him and recommended Torricelli. Galileo, blind and ill at the time, posed the problem to Torricelli on how to make a vacuum that, at the time, scientists could not believe existed.

On Galileo's death on January 8, 1642, Torricelli was appointed mathematician to the grand duke of Tuscany and Galileo's successor as professor of mathematics to the Florentine Academy at the request of Ferdinando II dei Medici. With a good salary and lodging in the Medici palace, he began his work.

Pumpmakers of the grand duke had a problem trying to raise water to a height of 40 inches or more and found that 32 feet was the upper limit to which they could raise water in a suction pump. Torricelli discovered that, because of atmospheric pressure, water will not rise above 33 feet in a suction pump. In 1643, he experimented by filling three-foot-long glass tubes with mercury, covering one end, and inverted each into an open bowl of mercury. He noticed that the mercury did not completely empty the tube and stayed in a column about 30 inches (76 centimeters) high above the mercury in the bowl, with an empty space above, although it varied day by day. This vacuum is known as the Torricellian vacuum.

In 1644, Torricelli presented a paper to explain his tube experiments, his "argento vivo" (mercury). In effect, he had discovered air pressure and the creation of the barometer, a device that measures air pressure. He explained the results as the heavyweight effect of being at the bottom of a sea of air, miles high, pressing down on the mercury in the bowl, balancing the weight of the mercury held up in the tube.

Because no one had ever created a sustained vacuum before the Torricellian vacuum, this finding was an important scientific discovery. Torricelli also observed that light and magnetism were conducted through the vacuum as well as through the atmosphere. However, he stayed away from the vacuum debate. Further experiments on whether animals could survive in a vacuum failed. It was several years before the scientific community finally accepted Torricelli's vacuum, based on experiments by later scientists, especially Robert Boyle. His work inspired other scientists, such as French mathematician and physicist, Blaise Pascal. He used a Torricellian tube and determined that thinner air at a high altitude has a lower air pressure, he also saw that a drop in air pressure often precedes rain or snow. Torricelli's experiments also were instrumental in German physicist Otto von Guericke's success in making a practical air pump.

Torricelli's discovery that water will not rise above 33 feet in a suction pump because of atmospheric pressure gives him the honor of being the father of the principles of hydromechanics. Although he is best known for being the inventor of the barometer (the word coined by Boyle in 1665, not by Torricelli), he worked on a variety of projects in math, fluids, and motion. He improved both telescopes and microscopes, being an excellent lens grinder, and published a number of papers in mathematics. In 1644, he published *Geometric Works*, which described his conclusions on the motion of fluids and projectiles. One of the units used to measure pressure, a torr, is named in honor of Torricelli. He died young, at age 31, from an unknown illness.

U

Uman, Martin A.
(1936–)
American
Electrical Engineer, Physicist

Martin Uman was born on July 3, 1936, in Tampa, Florida, to Morrice and Edith Uman, both attorneys. As a youth, Uman's interests were in tennis while he attended Gorrie Elementary, Woodrow Wilson Junior High, and H. B. Plant High School, all in Tampa. Uman received his bachelor's degree (1957), master's degree (1959), and Ph.D. (1961) in electrical engineering from Princeton University.

Uman's first position after graduate school was associate professor of electrical engineering at the University of Arizona, where he taught and did research in electromagnetics and gaseous electronics. It was there that he first became interested in the physics of lightning. In 1965, he joined the staff of the Westinghouse Research Laboratories in Pittsburgh as a fellow physicist and studied the physical and electromagnetic aspects of lightning and long laboratory sparks. Uman became a professor at the University of Florida in 1971. He has been director of the UF Lightning Research Laboratory since 1972. He cofounded and served as president of Lightning Location and Protection, Inc. (LLP), from 1975 to 1983. LLP, now a division of Global Atmospherics, is the

Martin A. Uman. Uman is generally acknowledged to be one of the world's leading authorities on lightning. He is probably best known for his work in lightning modeling: the application of electromagnetic field theory to the description of various lightning processes. *(Courtesy of Martin A. Uman)*

world leader in the sale of lightning locating equipment. Since 1991, Uman has been professor and chair of the department of electrical and computer engineering at the University of Florida.

Uman is generally acknowledged to be one of the world's leading authorities on lightning. He is probably best known for his work in lightning modeling: the application of electromagnetic field theory to the description of various lightning processes. That work, in addition to providing a better understanding of lightning in general, has had a number of important practical applications and spin-offs, the most notable being the LLP lightning locating system and the redefinition of several important lightning characteristics relative to hazard protection. Uman has published three books on the subject of lightning, with a fourth book now in press (coauthored with Vladimir A. RAKOV), as well as a book on plasma physics and 10 book chapters and encyclopedia articles on lightning. He has also published more than 150 papers in reviewed journals and more than 160 in other articles and reports. He holds five patents, four in the area of lightning detection and location.

Uman was the recipient of the 2001 American Geophysical Union John Adam Fleming Medal for original research and technical leadership in geomagnetism, atmospheric electricity, space science, aeronomy, and related sciences for "outstanding contribution to the description and understanding of electricity and magnetism of the earth and its atmosphere." He was awarded the Heinrich Hertz Medal in 1996 by the Institute of Electrical and Electronic Engineers (IEEE) for "outstanding contributions to the understanding of lightning electromagnetics and its application to lightning detection and protection." He was named the Florida Scientist of the Year by the Florida Academy of Sciences for 1990 and the 1988–89 University of Florida Teacher-Scholar of the Year, the highest UF faculty award. He is a Fellow of the Institute of Electrical and Electronic Engineers (IEEE), the American Geophysical Union, and the American Meteorological Society.

Uman is currently working on lightning protection of power lines for Florida Power and Light, lightning protection of aircraft for the Federal Aviation Administration (FAA), and basic lightning physics for the National Science Foundation.

V

Vonnegut, Bernard
(1914–1997)
American
Chemist, Meteorologist

Bernard Vonnegut was born in Indianapolis, Indiana, to Kurt Vonnegut, an architect, and Edith (Lieber) Vonnegut. He grew up with his brother, famed writer Kurt Vonnegut, and his sister Alice in Indianapolis. Vonnegut graduated from the Massachusetts Institute of Technology (MIT) in 1936 and continued at the school, receiving his Ph.D. in physical chemistry three years later. He worked for industrial laboratories (Hartford Empire Co., 1939–41) and as a MIT research associate (1941–45) before joining the labs of General Electric in Schenectady, New York, in 1945.

Vonnegut's first year at GE found him teamed up with Nobel Prize winner, Irving LANGMUIR, and Vincent SCHAEFER in a project for the U.S. Army Signal Corps, involving turning fog and clouds into rain and clearing overcast skies. Langmuir and Schaefer were working on aerosol-related war research until 1943, when they turned their attention to the investigation of the properties of supercooled clouds as part of a study of precipitation static.

Vonnegut worked under Langmuir at the General Electric Company from 1945 to 1952.

While there, he improved a method that had been pioneered by Schaefer, for artificially inducing rainfall, known as cloud seeding by using silver iodide as a cloud-seeding agent. Schaefer discovered in 1946 that a tiny grain of dry ice produced millions of ice crystals when dropped into a cloud of water droplets below the freezing point. This experiment was created using a special cloud chamber created by Schaefer called the cold box, a GE home freezer with a black velvet lining and a viewing light.

Vonnegut found that silver iodide had better results in nucleating clouds than did dry ice. On November 14, 1946, he deposited particles of silver iodide in a refrigerated chest of supercooled, liquid water droplets. The result was the formation of an abundance of ice crystals. No other nucleating process has been found that rivals the technique for making rain efficiently. It is still the preferred method used by commercial rainmakers to control drought in parts of the western United States and in more than 60 countries around the world.

After GE's legal department canceled further studies in February 1947, the U.S. Army Signal Corps, the Office of Naval Research, and the air force took over the idea and created Project Cirrus to transform supercooled water droplets into ice crystals, the main ingredient of cirrus clouds. The project moved to New Mexico in 1948 with

Langmuir, Schaefer, and Vonnegut working between there and Schenectady.

Vonnegut teamed up with Charles Moore, who helped create the Irving Langmuir Laboratory in New Mexico, and continued his research on thunderstorms and lightning. Vonnegut and colleagues developed lightning-observation equipment, produced controversial theories about the property of lightning and rain, and personally trained several space-shuttle crews in how to observe lightning while in space for scientific purposes.

Vonnegut left GE in 1952 and moved on to the A. D. Little Company, a scientific think tank in Cambridge, Massachusetts, until 1967, when he became professor of atmospheric science at New York State University in Albany, New York. He retired there in 1985, though he maintained an office and a workspace to his last days.

While at Albany, he was a popular teacher, earning the Distinguished Professor Award in 1983, which is the university's highest honor. In 1996 he was given the Award for Outstanding Contribution to Advancement of Applied Meteorology by the American Meteorological Society "for his pioneering discoveries of artificial techniques for nucleation of ice crystals which have continued to provide the basis for weather modification." The following year, the Weather Modification Association gave him the Vincent J. Schaefer Award for "scientific and technical discoveries that have constituted a major contribution to the advancement of weather modification."

Vonnegut continued studying meteorological phenomena, becoming an expert in such phenomena as lightning bolts and their role in precipitation and tornado formation, cloud electrification, and the effects of updrafts/downdrafts in thunderstorms. The holder of 28 patents, he was also the author of more than 160 articles and reports and contributed chapters to books and encyclopedias.

Vonnegut died of cancer on April 25, 1997, and is survived by his brother, esteemed novelist Kurt Vonnegut, five sons, and five grandchildren. His wife, Lois Bowler Vonnegut, died in 1972.

Wang, Daohong
(1964–   )
Chinese
Physicist, Electrical Engineer

Daohong Wang was born in Hubei, China, the son of middle-school teachers, Youfai Wang and Tianzheng Shang. Wang spent his early years in school in Hubei Province and on a farm and then attended Hubei Teacher College, where he received a bachelor of science degree in physics in 1984. In 1987, he received a master's degree in atmospheric physics from the Chinese Academy of Science, and in 1995, he earned a Ph.D. in electrical engineering from Osaka University of Japan.

After he received his master's degree, he began research work in Lanzhou Plateau Institute of Atmospheric Physics, at the Chinese Academy of Science. He worked there as an assistant researcher for nearly four years. His focus was to study the characteristics of plateau lightning, and he successfully triggered positive polarity lightning (discharges lower positive charge to ground) at the plateau area in the summer seasons.

Wang's own research was on rocket-triggered lightning, and he published his first paper, "The criterion for propagation of the positive streamer and its application," in the *Journal Acta Meteorologica Sinica* in 1989. His contribution in the

Daohong Wang. His contribution in the field was the discovery of electrical characteristics of rocket-triggered lightning, lightning attachment process, and leader pulse propagation characteristics. These discoveries have improved the knowledge of lightning physics. *(Courtesy of Daohong Wang)*

field was discovering electrical characteristics of rocket-triggered lightning, lightning attachment process, and leader pulse propagation characteristics. These discoveries have improved the knowledge of lightning physics. Discovery of the lightning leader pulse characteristics was a major accomplishment.

Wang received the Natural Science Award from the Chinese Academy of Science in 1991 and a Japanese Notari Scholarship. He married Dongxing Li in 1991 and they have one child. He continues his work in the department of human and information systems at Gifu University in Japan. His present study is of the natural-lightning attachment process, which is very important in lightning protection theory. He is also studying lightning protection of new energy systems, such as windmill generators and photovoltaic systems and looking at control issues of new energy systems.

⊠ Wang, Mingxing
(1944–   )
Chinese
Atmospheric scientist

Mingxing Wang was born in the Shandong Province, China, on January 18, 1944, to Wang Zhenli, a peasant, and mother Feng Guixiang. The Shandong Province is located at the estuary of the Yellow River and is China's second-most-populous province with 88.8 million people.

As a boy, Wang went to school in his small hometown village of Dalizhuang in Laixi, County of Shandong, and had interests of becoming a medical doctor or engineer. As a boy, he played sports, especially table tennis, and played for his school team (1956–62). He was also a member of his university team at Shandong University (1962–66), where he received a B.A. in physics (1968). Before he moved on to graduate school, he married Niu Xuhui in 1971. They have a son. In 1977, he began studies at Oxford University in the department of atmospheric physics (1977–78) in an advanced study program under Sir John Houghton.

Wang began his career as a researcher, conducting laboratory experiments on the infrared absorption of carbon dioxide for applications in remote sounding of the atmosphere from satellites. He also conducted field measurements of aerosol composition in North China in cooperation with. J. W. Winchester of Florida State University in 1980.

His first paper, in Chinese, entitled "Infrared Absorption by the 15 Micron Meter Band of Carbon Dioxide," was published in 1974 in the *Chinese Institute of Atmospheric Physics Annual Report*, published by Science Press Beijing. It was a summary report on the laboratory experiments that he conducted with Winchester in Florida.

Wang has made several important discoveries and observations in atmospheric science and in particular on the issue of acid rain and for the understanding of global warming, its causes, and future trends.

He focused his study on the chemistry of aerosols over remote areas and coal-fired cities in China. He has studied in detail the bimode-size distribution of the elements silicon (Si) and sulfur (S) in atmospheric aerosols over North China and nonacid rain in relation to chlorine deficits in atmospheric aerosols. This work is significant for the study of formation and control of acid rain and the study of heterogeneous atmospheric chemistry.

Wang also developed time to the study of methane emission from rice fields. He made significant contributions in the knowledge of the characteristics of the temperate and spatial variations of the emission rates and the controlling factors and in the understanding of the mechanism of methane formation in paddy soil and the transport and oxidation of methane in rice ecosystems. Among other new findings, he was the first to discover and investigate the diel variation of methane emissions from rice fields, which is caused by variations of the methane transport efficiency from soil to the atmosphere, not the variation of methane production in the soil. Furthermore, he found total methane emissions from Chinese and global rice fields that are only 10 and 30 Tg per year respectively (1 Tg = 1 million tons), one-fourth of an early estimate by other scientists that indicated that rice fields were not an important source of atmospheric methane. These findings are significant for the study of the concentration increase trend of atmospheric methane, which is an important greenhouse gas, and for the mitigation option of methane emission from rice fields.

He has been recognized for his work by receiving the China National Award for Advances in Science and Technology and an award for natural science from the Chinese Academy of Sciences. Wang has also published many papers and three important books: *Global Warming* (1996), *Atmospheric Chemistry* (1991), and *Methane Emission from Chinese Rice Field* (2001). He currently is professor of atmospheric chemistry at the Chinese Academy of Sciences, Institute of Atmospheric Physics (IAP), Beijing, where he currently is investigating biogenic (produced by living organisms or biological processes) emissions in China. The IAP is China's highest academic organization in the basic research of atmospheric sciences, and he has served as its director from 1997 to 2001.

⊠ Wegener, Alfred Lothar
(1880–1930)
German
Geophysicist, Meteorologist, Climatologist

Alfred Wegener was born in Berlin on November 1, 1880, the son of a minister who ran an orphanage, and obtained his doctorate in planetary astronomy in 1904 at the University of Berlin. In 1905, Wegener took a job at the Royal Prussian Aeronautical Observatory near Berlin, studying the upper atmosphere with kites and balloons. Wegener was an expert balloonist, as proved the following year when he and his brother Kurt set a world record of 52 hours straight in an international balloon contest.

He was invited to go on an expedition to Greenland as the official meteorologist, from 1906 to August 1908, under the leadership of Ludwig Mylius Erichsen. The purpose of this "Denmark" expedition of 28 men was to study polar air circulation and to explore the northeast coast, among other things. Wegener became the first to use kites and tethered balloons to study the polar atmosphere. However, Erichsen and two others (Jorgen Bronland and Hoegh-Hagen) died on the trip. Ironically, it would be Wegener's fate more than 25 years later. Returning to Germany in 1909, Wegener accepted a post as tutor at the University of Marburg, taking time to visit Greenland again in 1912–13. He taught at Marburg until 1919.

The idea that continents were connected was not new. In 1858, Antonio Snider-Pellegrini suggested that continents were linked during the Pennsylvanian period some 325 million years ago because certain Pennsylvanian plant fossils in Europe and North America were similar. In 1885, Australian geologist Edward Seuss saw similarities between plant fossils from South America, India, Australia, Africa, and Antarctica. He coined the word *Gondwanaland* for this proposed ancient supercontinent that was made up of these five landmasses. Finally, in 1910, American physicist Frank B. Taylor proposed the concept of continental drift to explain formation of mountain belts in a published paper, but it received no reaction.

In 1911, Wegener collected his meteorology lectures and published them as a book, *The Thermodynamics of the Atmosphere*, which became a standard in Germany and garnered Wegener much acclaim. While at Marburg, he noticed the close fit between the coastlines of Africa and South America. His formulation of his theory of continental drift was the beginning of his search for paleontological, climatological, and geological evidence in support of his theory.

On January 6, 1912, at a meeting of the Geological Association in Frankfurt, he spoke about his ideas of "continental displacement" (continental drift), and told it again days later at meeting of the Society for the Advancement of Natural Science in Marburg. He also married the daughter of eminent climatologist Vladimir Koppen and then returned to Greenland, making the longest crossing of the ice cap ever made on foot, 750 miles of snow, and ice and nearly dying. His expedition became the first to overwinter on an ice cap on the northeast coast.

Wegener published the results of the data he collected on the polar trips, becoming a world expert on polar meteorology and glaciology, and he was the first to trace storm tracks over the ice cap. In 1914, he was drafted into the German army, was wounded, and served out the war in the army weather-forecasting service. While recuperating in a military hospital, he further developed his theory of continental drift that he published the following year as *Die Entstehung der Kontinente und Ozeane* (The origin of continents and oceans). Expanded versions of the book were published in 1920, 1922, and 1929. Wegener wrote that about 300 million years ago, the continents had formed a single mass, called Pangaea (Greek for "all the Earth") that split apart, and its pieces had been moving away from each other ever since. He was not the first to suggest that the continents had once been connected, but he was the first to present the evidence, although he was wrong in thinking that the continents moved by "plowing" into each other through the ocean floor. His theory was soundly rejected, although a few scientists did agree with his premise.

In 1921, Wegener strayed a bit from meteorology and finally published in the field that gave him his doctorate, astronomy. *Die Entstehung der Mondkrater* (The origin of the lunar craters) was his attempt to argue that the origin of lunar craters was the result of impact and not volcanic, as proposed by others. In 1924, he accepted a professorship in meteorology and geophysics at the University of Graz, in Austria. It was also the year that he published, with his father in law, V. Köppen, "Climate and Geological Pre-history."

He returned to the Greenland ice cap in 1930 with 21 people in an attempt to measure the thickness and climate of the ice cap. In November 1930, he died while returning from a rescue expedition that brought food to a party of his colleagues who were camped in the middle of the Greenland ice cap. His body was not found until May 12, 1931, but his friends allowed him to rest forever in the area that he loved.

The theory of continental drift continued to be controversial for many years, but by the 1950s and 1960s, plate tectonics was all but an accepted fact and taught in schools. Today, we know that both continents and ocean floor float as solid plates on underlying rock that behaves like a viscous fluid, due to being under such tremendous heat and pressure. Wegener never lived to see his theory proved. Had he lived, most scientists believe he would have been the champion of present-day plate tectonics.

⊠ **Wielicki, Bruce Anthony**
(1952–   )
American
Oceanographer

Bruce Wielicki was born on October 13, 1952, in Milwaukee, Wisconsin, to Anthony Francis Wielicki, a mechanical engineer, and his mother Penelope. As a youngster, while in Sacred Heart Elementary (St. Martins, Wisconsin) and New Berlin High School (New Berlin, Wisconsin), he had interests in model building, reading, electric trains, football, and an early fascination with oceanography. It was his high school counselor who instilled the fact that being involved in oceanographic research would require going all the way to a Ph.D. In between his first job, working with a gravel-pit logging truck, loading gravel for a local freeway project, and summer jobs in a tool and Koss Headphones warehouse, he consulted many books that recommended earning a good physics and math degree in undergraduate school, followed by oceanography as a field in graduate school.

He attended the University of Wisconsin at Madison and received his bachelor of science degree in applied mathematics and engineering physics in 1974. As an undergraduate, he worked as an assistant to a graduate student who was studying currents in Lake Superior. Wielicki manually digitized aerial photographs of floating poster boards that drifted with the currents. He received his Ph.D. in physical oceanography from Scripps Institution of Oceanography at the University of California in San Diego in 1990.

Wielicki's research has focused for more than 20 years on clouds and their role in the Earth's radiative energy balance. The Sun, which heats up the Earth until it is hot enough to emit as much energy as it receives from the Sun, makes it radiatively balanced. His first research project was evaluating the accuracy of satellite remote sensing of cloud heights and cloud amounts, and he published his first paper in 1981 with J. A. Coakley entitled "Cloud Retrieval Using Infrared Sounder Data" in the *Journal of Applied Meteorology*. This paper used radiative transfer theory and the new computational capability that was becoming available to evaluate the ability of spaceborne sensors to determine remotely the Earth's global cloud cover and height.

He currently serves as lead and principal investigator of the Clouds and the Earth's Radiant Energy System (CERES) Instrument investigation, an international science and engineering team. CERES is part of NASA's Earth Observing System, designed to explore the Earth's climate system and narrow the uncertainties in predicting future climate change. Wielicki is also principal investigator of the CERES Interdisciplinary Science investigation. The CERES mission is designed to provide the first integrated measurements of cloud properties and radiative fluxes from the surface to the top of the atmosphere by synergistically combining simultaneous coarse-resolution broadband-scanner radiance data with

narrow-band high-resolution cloud-imager estimates of physical cloud properties. Launch of the CERES instruments began in November 1997 on the Tropical Rainfall Measuring Mission (TRMM) and continued on EOS-Terra in December 1999 and the EOS-Aqua in May 2002.

Wielicki led an international team of climate modelers and observationalists who compared 22 years of global satellite radiation data with current global climate models. The results found surprising variations in the tropical radiation fields during the last two decades. These variations were not reproduced in the best current climate models, suggesting that clouds continue to be a major problem in predicting future climate change.

Wielicki was also a coinvestigator on the Earth Radiation Budget Experiment (ERBE) and developed a new maximum-likelihood estimation (MLE) method for determining the cloud condition in each ERBE field of view. This method required only the ERBE data itself and enabled the development of the first estimates of cloud radiative forcing (CRF) in the climate system by distinguishing individual observations as clear, partly cloudy, mostly cloudy, or overcast. The term *cloud radiative forcing* refers to the effects that clouds have on both sunlight and heat in the atmosphere. This measurement became a standard of comparison for global climate models. The poor ability of global climate models to reproduce the ERBE cloud radiative forcing measurements was a key element in the designation of the role of clouds and radiation as the highest priority of the U.S. Global Change Research Program. Today, clouds remain the largest single uncertainty in predictions of future climate change.

Wielicki was a principal investigator on the First ISCCP Regional Experiment (FIRE). ISCCP is the International Satellite Cloud Climatology Experiment. He served as FIRE project scientist from 1987 to 1994. His research, using Landsat satellite data, provided the first definitive validation of the accuracy of satellite-derived cloud fractional coverage, and he developed new methods to derive cloud-cell-size distributions from space-based data. He demonstrated non-gaussian distributions of cloud optical depth present in broken boundary layer cloud fields, and he has shown the large bias that these distributions can cause in global climate model estimates of both solar and thermal infrared fluxes.

Throughout his career, Wielicki has pursued extensive theoretical radiative transfer studies of the effects on nonplanar cloud geometry on the calculation of radiative fluxes and has gathered space-based observations on the retrieval of cloud properties and top of atmosphere radiative fluxes. His research also determined the accuracy with which infrared sounder data can be used to sense cloud height and cloud emissivity remotely. He has also extended this work to examine the ability to combine AVHRR (The Advanced Very High Resolution Radiometer) and HIRS (The High Resolution Infrared Sounder) data to measure multilayered clouds.

He has been as an associate editor for *Journal of Climate*, and a member of the International Commission on Clouds and Precipitation. He currently serves on the American Geophysical Union (AGU) Committee on Global Environmental Change, on the U.S. CLIVAR (Climate Variability) Executive Science Steering Committee, and on the international World Climate Research Program's GEWEX Radiation Panel. GEWEX is the Global Energy and Water Cycle Experiment. In 1979, Wielicki married Barbara S. Stone. They have two children.

Wielicki has published more than 50 authored or coauthored journal papers and a similar number of conference papers. He received the NASA Outstanding Leadership Medal in 2002, the American Meteorological Society Henry G. Houghton Award in 1995, and the NASA Exceptional Scientific Achievement Award in 1992. He currently is a member of the American Meteorological Society and American Geophysical Union.

Wielicki's most significant contribution is providing the science community with insights into the role of clouds in global climate and of cloud's effect on the radiation energy balance of the Earth.

⊠ Williams, Earle Rolfe
(1951–)
American
Geophysicist

Earle Williams was born on December 20, 1951, in South Bend, Indiana, to Warner Williams, an artist and sculptor, and Jean Aber Williams, a calligrapher and painter. He attended the Culver Community Schools in Culver, Indiana, from grades 1 to 7 and Culver Military Academy from grades 8 to 12. As a youth, he had a longstanding interest in the natural world, canoeing, and long-distance river trips and a fascination with weather and the severe thunderstorms so prevalent in the Midwest. He is also a direct descendant from Pocahontas, whose husband's name was John Rolfe. After graduation from Culver, he attended Swarthmore College in Pennsylvania and in 1974 received a B.A. in physics with high honors. Following summer jobs measuring salinity in estuaries all over Cape Cod, Massachusetts, and dabbling in experimental nuclear physics at Brookhaven National Laboratory, he decided on geophysics as a career and entered graduate school at Massachusetts Institute of Technology (MIT). There he developed interests in cloud physics, thunderstorm electrification, radar meteorology, breakdown phenomena, and tropical meteorology. He received a Ph.D. in geophysics in 1981, the year after he married Kathleen Moore Bell. They have two children.

Williams has participated in numerous field programs to study thunderstorms throughout the world. "Radar tests of the precipitation hypothesis for thunderstorm electrification" was his first publication, appearing in the *Journal of Geophysical Research* in 1983. This study was aimed at measurements of changes in the fall speeds of raindrops and small hailstones during lightning discharges, using a vertically pointing Doppler radar.

Williams has contributed to the field of meteorology in three areas. He discovered in 1991 that the global electrical circuit is responsive to atmospheric temperature on a variety of time scales. This has demonstrated that the global electrical circuit is a natural and inexpensive framework for monitoring global change and from only a single measurement station. In 1995, he found, with student Dennis Boccippio and research colleague Walter Lyons, that the Earth's Schumann resonances are strongly excited by giant lightning discharges that also make sprites in the mesosphere—a new kind of lightning. This is significant because it linked a large body of ELF (extremely low-frequency: 3 Hz-1500 Hz) work with a new body of work on upper atmosphere discharges.

Finally, he discovered in 2000, with student Akash Patel, that the global five-day planetary wave is detectable over Africa, with Schumann resonance observations of large lightning transients. Here, the robustness of the global five-day wave was demonstrated, as was its possible role in the large-scale circulation of the atmosphere. The Schumann resonances are global-scale electromagnetic waves that are continuously excited by worldwide lightning discharges. The waves are trapped in the spherical cavity formed by the conductive Earth and the conductive ionosphere. The wavelength for the fundamental resonance ($\sim$8 Hz) is equal to the circumference of the Earth.

Williams has received numerous research grants from the U.S. National Science Foundation and NASA and has benefited in international cooperation from support from the U.S.–Hungary Joint Science Foundation and U.S.–Israel Binational Science Foundation. As the author of more than 100 articles, papers, and conference proceedings, he has received numerous academic awards and is a member of the American Geophysical Union and the American Meteorological Society. He was secretary of the International Commission on Atmospheric Electricity for eight years.

He continues to work on global circuit responsiveness to worldwide weather as the principal research scientist in the department of civil and environmental engineering at MIT's Parsons Laboratory in Cambridge, Massachusetts, and on issues of aviation weather at the MIT Lincoln Laboratory.

⊠ Wyrtki, Klaus
(1925–)
German/American
Oceanographer

Klaus Wyrtki was born in Tarnowtiz, Germany, on February 7, 1925. His father died when he was four, and he was raised by his mother. When he was 14, he dreamed of becoming a naval architect, designing aircraft carriers, and he enlisted in the navy hoping to accomplish that goal. After World War II, he changed his plans and enrolled in the University of Marburg to study mathematics and physics. Reading books, he discovered meteorology and oceanography and moved on to the University of Kiel, Germany, to study oceanography (1948–50). He received his doctor of science (magna cum laude) under the supervision of George Wust. For his Ph.D. study, Wust told Wyrtki to go to the German Hydrographic Institute and borrow an instrument that measures turbidity in the ocean and "just take the instrument and go out to sea and measure more often than anybody has measured with it. And you will find something new."

He worked from 1950 to 1951 at the German Hydrographic Institute in Hamburg and from 1951 to 1954 for the German Research Council on a postdoctoral research fellowship at the University of Kiel. In 1954, he married and later fathered two children. In 1954, he became the

Klaus Wyrtki. Wyrtki's legacy is that he not only conducted the first studies of circulation in the Hawaiian archipelago more than 30 years ago but also conducted some of the first work on the El Niño phenomenon. *(Courtesy of Klaus Wyrtki)*

head of the Institute of Marine Research at Djakarta, Indonesia. Here, he had a 200-ton research vessel called the *Samudera*. With very little instrumentation, he conducted "a few surveys with Nansen bottles down to a few hundred meters but could not reach the deep sea basins in Indonesia because of a lack of a long wire, and that restricted us to the surface layers." He found a great deal of data on the area that had never been analyzed and wrote a book on it, *Physical Oceanography of the Southeast Asian Waters*, published by the University of California. The book remained a valuable reference for decades.

During this time, Wyrtki discovered the Indonesian through-flow. The Indonesian through-flow affects the circulation around Australia and in the Indian Ocean; it increases surface temperatures in the eastern Indian Ocean and forms the surface link of the ocean conveyor belt.

He left Indonesia in 1957 and became a senior research officer, later principal research officer, of the Commonwealth Scientific and Industrial Research Organization, Division of Fisheries and Oceanography, in Sydney, Australia. Here he wrote papers on thermocline circulation and on the oxygen minima in the oceans. It was here that it became clear to him that "vertical movements are the main links in ocean circulation—like the Antarctic upwelling, like the vertical movements in the deep ocean basins that must bring slowly up water to the surface and are counteracted by vertical diffusion."

From 1961 to 1964 Wyrtki was a research oceanographer at the University of California's Scripps Institution of Oceanography, and he did research on the eastern tropical Pacific and the upwelling off Peru. In 1964, he became professor of oceanography at the University of Hawaii, where he remained until he retired in 1995. There, he started a project that was dear to his heart, investigating the circulation of the Indian Ocean. He discovered an equatorial jet in the Indian Ocean that moves water from west to east twice a year in response to the monsoons and is related to vertical movements of the thermocline at each end. He produced the *Oceanographic Atlas of the International Indian Ocean Expedition*, a book that was largely made by computer data and mapping and was published in 1972.

Wyrtki's most outstanding contribution was his explanation of El Niño. In 1975, he claimed that a collapse of the trade winds along the equator would trigger a massive surge of water, a Kelvin wave, that moves warm water from the western Pacific along the equator to the eastern Pacific. He used wind observations from merchant ships, sea-level data from islands, and the displacement of the thermocline to make his case. He expanded the sea-level network over most islands in the Pacific and Indian Oceans and collected data to prove his theory during succeeding El Niño events. He explained El Niño cycles by demonstrating that the volume of the warm upper layer in the equatorial Pacific increases slowly between El Niño events and then rapidly decreases during El Niño, representing a heat relaxation of the ocean–atmosphere system. He was also the first to estimate the volume rate of equatorial upwelling in the Pacific Ocean.

Wyrtki was active in creating the Global Sea Level Network (GLOSS) and advocating the establishment of a global ocean-monitoring system. During this career, he produced 130 publications, and he is a member or Fellow of the American Geophysical Union, the American Meteorological Society, the Hawaiian Academy of Science, and the Oceanography Society. He has served as a chair or member of several committees and panels and was a frequent lecturer.

He has received many awards, including the Excellence in Research Award, University of Hawaii (1980); the Rosenstiel Award in Oceanographic Sciences, University of Miami (1981); the Maurice Ewing Medal, American Geophysical Union (1989); the Sverdrup Gold Medal, American Meteorological Society (1991); and the Albert Defant Medal, Deutsche Meteorologische Gesellschaft (1992). He became an American citizen in 1977.

In 2001, the *Shaman II*, a 57-foot longline fishing vessel donated to the University of Hawaii, was renamed for Wyrtki. It is used by faculty and students for coastal investigations. Wyrtki's legacy is that he not only conducted the first studies of circulation in the Hawaiian archipelago more than 30 years ago but also conducted some of the first work on the El Niño phenomenon. The Wyrtki Center at the University of Hawaii at Manoa is part of the NOAA Joint Institute for Marine and Atmospheric Research (JIMAR) and is in the School of Ocean and Earth Science and Technology (SOEST). The center is named to honor Wyrtki for his pioneering research on the El Niño/Southern Oscillation (ENSO) phenomena, including the earliest attempts to determine the predictability of ENSO events.

Z

⊠ Zajac, Bard Anton
(1972–)
American
Atmospheric Scientist

Bard Zajac was born on July 20, 1972, in Bethesda, Maryland, to Blair Zajac, an engineer for Boeing, and Barbara Zajac, a high school substitute teacher. Growing up in rural Issaquah, western Washington, he lived a typical life of outdoor activity and sports. He worked summers in landscaping and construction until he graduated in 1990.

He attended the University of Washington in Seattle and received his bachelor of science degree in physics in 1995. During summer months, Zajac worked as an intern at Columbia University (studying marine geology, specifically lithification of ocean sediment) in New York City, the New Mexico Institute of Mining & Technology (in the field of meteorology conducting field research on thunderstorms) in Socorro, and Colorado State University (in meteorology, conducting rainfall estimation in the tropics). He attended graduate school at Colorado State University in Fort Collins and received a master's degree in atmospheric sciences in 1998.

In 2001, he published his first paper with Steven A. Rutledge, his graduate advisor, entitled "Cloud-to-ground lightning activity in the contiguous United States from 1995 to 1999," in the

Bard Zajac. Zajac's work has demonstrated the value of cloud–ground lightning data in basic and applied meteorology. The data can be utilized in thunderstorm climatologies, storm case studies, and operational weather forecasting. *(Courtesy of Bard Zajac)*

Monthly Weather Review. This paper was based on his master's degree research and documented the spatial, annual, and diurnal distributions of cloud-to-ground (cg) lightning over the contiguous United States. The paper placed cloud–ground (cg) lightning observations, a relatively new data type, in the context of more-common data types such as rainfall, thunder reports, and severe weather reports. The paper also examined cg lightning activity over the north-central United States in detail. Several signals were identified over this region. These signals appear to identify an area over which a unique class of thunderstorms, often severe and producing hail and tornadoes, are relatively common. He has authored several conference papers and technical reports.

After earning his master's degree, Zajac took a job with the Cooperative Institute for Research in the Atmosphere (CIRA) at Colorado State University. Within CIRA, he worked with the National Weather Service, VISIT training group (VISIT stands for Virtual Institute for Satellite Integration Training). This group develops and presents training on remote sensing topics to NWS forecasters, using distance learning techniques. He has established a curriculum on forecasting with cg lightning data, which includes three 90-minute courses and associated web-based material. VISIT has been a successful program, earning him the Teacher of the Year Award from a related federal agency in 2000.

Zajac's work has demonstrated the value of cg lightning data in basic and applied meteorology. The data can be utilized in thunderstorm climatologies, storm case studies, and operational weather forecasting.

ENTRIES BY FIELD

ANTHROPOLOGY
Galton, Sir Francis

ARCHITECTURE
Alberti, Leon Battista

ASTRONOMY
Abbe, Cleveland
Ångström, Anders Jonas
Celsius, Anders
Galilei, Galileo
Hooke, Robert
Todd, Charles

ASTROPHYSICS
Milankovitch, Milutin

ATMOSPHERIC SCIENCE
Blanchard, Duncan C.
Cheng, Roger J.
Liou, Kuo-Nan
McIntyre, Michael
    Edgeworth
Salm, Jaan
Wang, Mingxing
Zajac, Bard Anton

BOTANY
Fritts, Harold Clark

CARPENTRY
Croll, James

CHEMISTRY
Arrhenius, Svante
    August
Dalton, John
Daniell, John Frederic
Howard, Luke
Langmuir, Irving
Solomon, Susan
Vonnegut, Bernard

CIVIL ENGINEERING
Bras, Rafael Luis

CLIMATOLOGY
Blodget, Lorin
Cane, Mark Alan
Sellers, Piers John
Symons, George James

CLOUD PHYSICS
Sinkevich, Andrei

DENDROCHRONOLOGY
Fritts, Harold Clark

ELECTRICAL ENGINEERING
Hale, Les
Mardiana, Redy

Rakov, Vladimir A.
Uman, Martin A.

ENGINEERING
Amontons, Guillaume
Milankovitch, Milutin
Stevenson, Thomas
Suomi, Verner

EXPLORATION
Galton, Sir Francis

GENERAL SCIENCE
Franklin, Benjamin
Leonardo da Vinci

GEOLOGY
Guyot, Arnold Henry

GEOPHYSICS
Chapman, Sydney
Sverdrup, Harald Ulrik
Wegener, Alfred
Williams, Earle Rolfe

HYDROGRAPHY
Maury, Matthew Fontaine

INSTRUMENT MAKING
Fahrenheit, Daniel Gabriel

MATHEMATICS
Archimedes
Bjerknes, Vilhelm Friman
    Koren
Chapman, Sydney
Espy, James Pollard
Galilei, Galileo
Richardson, Lewis Fry
Neumann, John Louis von
    (Johann)
Newton, Sir Isaac
Thomson, William
    (Lord Kelvin)
Torricelli, Evangelista

METEOROLOGY
Abbe, Cleveland
Anthes, Richard A.
Atlas, David
Bentley, Wilson Alwyn
Bjerknes, Jacob Aall Bonnevie
Bjerknes, Vilhelm Friman
    Koren
Buchan, Alexander
Buys Ballot, Christoph
    Hendrik Diederik
Charney, Jule Gregory
Daniell, John Frederic
Day, John
Emanuel, Kerry Andrew
Espy, James Pollard
Fitzroy, Robert
Friday, Jr., Elbert Walter (Joe)
Holle, Ronald L.
Howard, Luke
Kalnay, Eugenia Enriqueta
Katsaros, Kristina
King, Patrick
Kurihara, Yoshio
LeMone, Margaret Anne
Lilly, Douglas K.
Minnis, Patrick

Mintz, Yale
O'Brien, James Joseph Kevin
Rasmusson, Eugene Martin
Rossby, Carl-Gustaf Arvid
Schaefer, Vincent Joseph
Shaw, Sir William Napier
Simpson, Joanne Gerould
Simpson, Robert H.
Stensrud, David J.
Stolzenburg, Maribeth
Suomi, Verner
Teisserenc de Bort, Léon
    Philippe
Vonnegut, Bernard
Wegener, Alfred

NATURAL HISTORY
Réaumur, René-Antoine
    Ferchault de

OCEANOGRAPHY
Buchan, Alexander
Cane, Mark Alan
Franklin, Benjamin
Godfrey, John Stuart
Maury, Matthew
    Fontaine
Philander, S. George H.
Stommel, Henry Melson
Sverdrup, Harald Ulrik
Wielicki, Bruce Anthony
Wyrtki, Klaus

PHILOSOPHY
Alberti, Leon Battista
Aristotle
Hadley, George
Réaumur, René-Antoine
    Ferchault de

PHOTOGRAPHY
Bentley, Wilson Alwyn

PHYSICS
Amontons, Guillaume
Anisimov, Sergey Vasilyevich
Ångström, Anders Jonas
Arrhenius, Svante August
Beeckman, Isaac
Boltzmann, Ludwig Eduard
Brook, Marx
Celsius, Anders
Coriolis, Gaspard-Gustave de
Croll, James
Fahrenheit, Daniel Gabriel
Franklin, Benjamin
Hadley, George
Handel, Peter Herwig
Harrison, R. Giles
Henry, Joseph
Hertz, Heinrich Rudolph
Hooke, Robert
MacCready, Paul Beattie
Mach, Douglas Michael
Maxwell, James Clerk
McCormick, Michael Patrick
McIntyre, Michael
    Edgeworth
Mitseva-Nikolova, Rumyana
    Petrova
Richard, Philippe
Sentman, Davis Daniel
Shaw, Sir William Napier
Strutt, John William (Lord
    Rayleigh)
Surtees, Antony J.
Swanson, Brian Douglas
Thomson, William (Lord
    Kelvin)
Tierney (née Morris), Heidi
    Ellen
Tinsley, Brian Alfred
Todd, Charles
Torricelli, Evangelista
Wang, Daohong

Entries by Country of Birth

ARGENTINA
Kalnay, Eugenia Enriqueta

AUSTRIA
Boltzmann, Ludwig Eduard

AUSTRIA-HUNGARY
Milankovitch, Milutin
(Serbian)

AUSTRALIA
McIntyre, Michael
Edgeworth
Surtees, Antony J.

BULGARIA
Mitseva-Nikolova, Rumyana
Petrova

CANADA
King, Patrick

CHINA
Cheng, Roger J.
Wang, Daohong
Wang, Mingxing

ESTONIA
Salm, Jaan

FRANCE
Amontons, Guillaume
Coriolis, Gaspard-Gustave de
Réaumur, René-Antoine
Ferchault de
Richard, Philippe

GERMANY
Fahrenheit, Daniel Gabriel
Hertz, Heinrich Rudolph
Wegener, Alfred
Wyrtki, Klaus

GREAT BRITAIN

England
Chapman, Sydney
Dalton, John
Daniell, John Frederic
Fitzroy, Robert
Galton, Sir Francis
Godfrey, John Stuart
Hadley, George
Hooke, Robert
Howard, Luke
John Sellers, Piers
Richardson, Lewis Fry
Shaw, Sir William Napier
Sir Isaac Newton
Strutt, John William (Lord
Rayleigh)

Symons, George James
Todd, Charles

Scotland
Buchan, Alexander
Croll, James
Maxwell, James Clerk
Stevenson, Thomas

GREECE
Archimedes
Aristotle

HUNGARY
Neumann, John Louis von
(Johann)

Austria-Hungary
Milankovitch, Milutin
(Serbian)

INDONESIA
Mardiana, Redy

IRELAND
Thomson, William (Lord
Kelvin)

ITALY
Alberti, Leon Battista
Galilei, Galileo
Leonardo da Vinci

KAZAKHSTAN
Rakov, Vladimir A.

KOREA
Kurihara, Yoshio

NETHERLANDS
Beeckman, Isaac
Buys Ballot, Christoph
    Hendrik Diederik

NEW ZEALAND
Tinsley, Brian Alfred

NORWAY
Bjerknes, Vilhelm Friman
    Koren
Sverdrup, Harald Ulrik

PUERTO RICO
Bras, Rafael Luis

ROMANIA
Handel, Peter Herwig

RUSSIA
Anisimov, Sergey Vasilyevich
Sinkevich, Andrei

SOUTH AFRICA
Philander, S. George H.

SWEDEN
Ångström, Anders Jonas
Arrhenius, Svante August
Bjerknes, Jacob Aall Bonnevie
Celsius, Anders
Katsaros, Kristina
Rossby, Carl-Gustaf Arvid

SWITZERLAND
Guyot, Arnold Henry

TAIWAN, REPUBLIC OF CHINA
Liou, Kuo-Nan

UNITED STATES
Abbe, Cleveland
Anthes, Richard A.
Atlas, David
Bentley, Wilson Alwyn
Blanchard, Duncan C.
Blodget, Lorin
Brook, Marx
Cane, Mark Alan
Charney, Jule Gregory
Day, John
Emanuel, Kerry Andrew
Espy, James Pollard
Franklin, Benjamin
Friday, Jr., Elbert Walter
    (Joe)
Fritts, Harold Clark
Hale, Les

Henry, Joseph
Holle, Ronald L.
Langmuir, Irving
LeMone, Margaret Anne
Lilly, Douglas K.
MacCready, Paul Beattie
Mach, Douglas Michael
Maury, Matthew Fontaine
McCormick, Michael
    Patrick
Minnis, Patrick
Mintz, Yale
O'Brien, James Joseph
    Kevin
Rasmusson, Eugene Martin
Schaefer, Vincent Joseph
Sentman, Davis Daniel
Simpson, Joanne Gerould
Simpson, Robert H.
Solomon, Susan
Stensrud, David J.
Stolzenburg, Maribeth
Stommel, Henry Melson
Suomi, Verner
Swanson, Brian Douglas
Tierney (née Morris), Heidi
    Ellen
Uman, Martin A.
Vonnegut, Bernard
Wielicki, Bruce Anthony
Williams, Earle Rolfe
Zajac, Bard Anton

ENTRIES BY COUNTRY OF MAJOR SCIENTIFIC ACTIVITY

AUSTRIA
Boltzmann, Ludwig Eduard

AUSTRIA-HUNGRY (SERBIA)
Milankovitch, Milutin

AUSTRALIA
Godfrey, John Stuart
McIntyre, Michael Edgeworth
Todd, Charles

BULGARIA
Mitseva-Nikolova, Rumyana
    Petrova

CANADA
King, Patrick

CHINA
Cheng, Roger J.
Wang, Mingxing

ENGLAND

United Kingdom
Chapman, Sydney
Dalton, John
Daniell, John Frederic
Fitzroy, Robert
Galton, Sir Francis
Hadley, George

Hooke, Robert
Howard, Luke
Richardson, Lewis Fry
Shaw, Sir William Napier
Sir Isaac Newton
Strutt, John William (Lord
    Rayleigh)
Symons, George James

ESTONIA
Salm, Jaan

FRANCE
Amontons, Guillaume
Coriolis, Gaspard-Gustave de
Réaumur, René-Antoine
    Ferchault de
Richard, Philippe

GERMANY
Fahrenheit, Daniel Gabriel
Hertz, Heinrich Rudolph

GREECE
Archimedes
Aristotle

GREENLAND
Wegener, Alfred

INDONESIA
Mardiana, Redy

ITALY
Alberti, Leon Battista
Galilei, Galileo
Leonardo da Vinci

JAPAN
Kurihara, Yoshio
Wang, Daohong

NETHERLANDS
Beeckman, Isaac
Buys Ballot, Christoph
    Hendrik Diederik

NORWAY
Bjerknes, Vilhelm Friman
    Koren
Sverdrup, Harald Ulrik

RUSSIA
Anisimov, Sergey Vasilyevich
Sinkevich, Andrei

SCOTLAND
Buchan, Alexander
Croll, James
Maxwell, James Clerk

Stevenson, Thomas
Thomson, William (Lord Kelvin)

SWEDEN
Ångström, Anders Jonas
Arrhenius, Svante August
Bjerknes, Jacob Aall Bonnevie
Celsius, Anders

UNITED STATES
Abbe, Cleveland
Anthes, Richard A.
Atlas, David
Bentley, Wilson Alwyn
Bjerknes, Jacob Aall Bonnevie
Blanchard, Duncan C.
Blodget, Lorin
Bras, Rafael Luis
Brook, Marx
Cane, Mark Alan
Charney, Jule Gregory
Day, John
Emanuel, Kerry Andrew

Espy, James Pollard
Franklin, Benjamin
Friday, Jr., Elbert Walter (Joe)
Fritts, Harold Clark
Guyot, Arnold Henry
Hale, Les
Handel, Peter Herwig
Henry, Joseph
Holle, Ronald L.
Kalnay, Eugenia Enriqueta
Katsaros, Kristina
Langmuir, Irving
LeMone, Margaret Anne
Lilly, Douglas K.
Liou, Kuo-Nan
MacCready, Paul Beattie
Mach, Douglas Michael
Maury, Matthew Fontaine
McCormick, Michael Patrick
Minnis, Patrick
Mintz, Yale
Neumann, John Louis von (Johann)
O'Brien, James Joseph Kevin

Philander, S. George H.
Rakov, Vladimir A.
Rasmusson, Eugene Martin
Rossby, Carl-Gustaf Arvid
Schaefer, Vincent Joseph
Sellers, Piers John
Sentman, Davis Daniel
Simpson, Joanne Gerould
Simpson, Robert H.
Solomon, Susan
Stensrud, David J.
Stolzenburg, Maribeth
Stommel, Henry Melson
Suomi, Verner
Surtees, Antony J.
Swanson, Brian Douglas
Tierney (née Morris), Heidi Ellen
Tinsley, Brian Alfred
Uman, Martin A.
Vonnegut, Bernard
Wielicki, Bruce Anthony
Williams, Earle Rolfe
Wyrtki, Klaus
Zajac, Bard Anton

1910–1919
Charney, Jule Gregory
Day, John
Mintz, Yale
Simpson, Robert H.
Suomi, Verner
Vonnegut, Bernard

1920–1929
Atlas, David
Blanchard, Duncan C.
Brook, Marx
Cheng, Roger J.
Fritts, Harold Clark
Lilly, Douglas K.
MacCready, Paul Beattie
Rasmusson, Eugene Martin
Simpson, Joanne Gerould
Stommel, Henry Melson
Wyrtki, Klaus

1930–1939
Friday, Jr., Elbert Walter (Joe)
Hale, Les
Handel, Peter Herwig

Kurihara, Yoshio
O'Brien, James Joseph Kevin
Salm, Jaan
Tinsley, Brian Alfred
Uman, Martin A.

1940–1949
Anthes, Richard A.
Cane, Mark Alan
Godfrey, John Stuart
Holle, Ronald L.
Kalnay, Eugenia Enriqueta
King, Patrick
LeMone, Margaret Anne
Liou, Kuo-Nan
McCormick, Michael Patrick
McIntyre, Michael Edgeworth
Mitseva-Nikolova, Rumyana
    Petrova
Philander, S. George H.
Sentman, Davis Daniel
Wang, Mingxing

1950–1959
Anisimov, Sergey Vasilyevich

Bras, Rafael Luis
Emanuel, Kerry Andrew
Mach, Douglas Michael
Minnis, Patrick
Rakov, Vladimir A.
Richard, Philippe
Sellers, Piers John
Sinkevich, Andrei
Solomon, Susan
Surtees, Antony J.
Swanson, Brian Douglas
Wielicki, Bruce Anthony
Williams, Earle Rolfe

1960–1969
Harrison, R. Giles
Mardiana, Redy
Stensrud, David J.
Stolzenburg, Maribeth
Wang, Daohong

1970–1979
Tierney (née Morris), Heidi
    Ellen
Zajac, Bard Anton

CHRONOLOGY

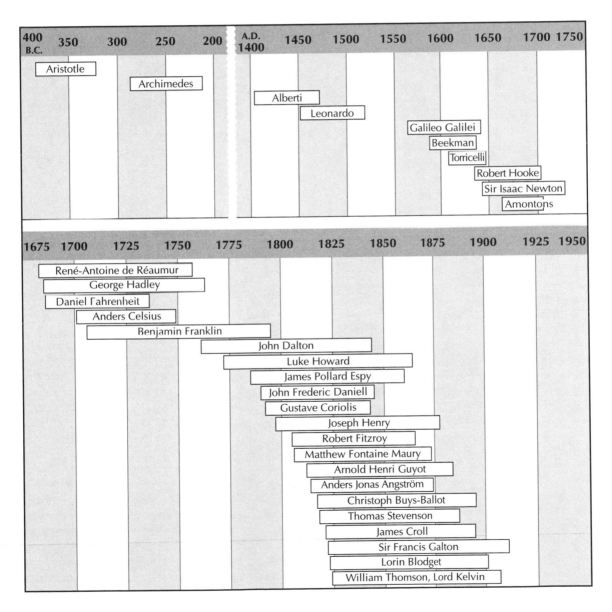

400 B.C.	350	300	250	200		A.D. 1400	1450	1500	1550	1600	1650	1700	1750

Aristotle

Archimedes

Alberti

Leonardo

Galileo Galilei

Beekman

Torricelli

Robert Hooke

Sir Isaac Newton

Amontons

1675	1700	1725	1750	1775	1800	1825	1850	1875	1900	1925	1950

René-Antoine de Réaumur

George Hadley

Daniel Fahrenheit

Anders Celsius

Benjamin Franklin

John Dalton

Luke Howard

James Pollard Espy

John Frederic Daniell

Gustave Coriolis

Joseph Henry

Robert Fitzroy

Matthew Fontaine Maury

Arnold Henri Guyot

Anders Jonas Angström

Christoph Buys-Ballot

Thomas Stevenson

James Croll

Sir Francis Galton

Lorin Blodget

William Thomson, Lord Kelvin

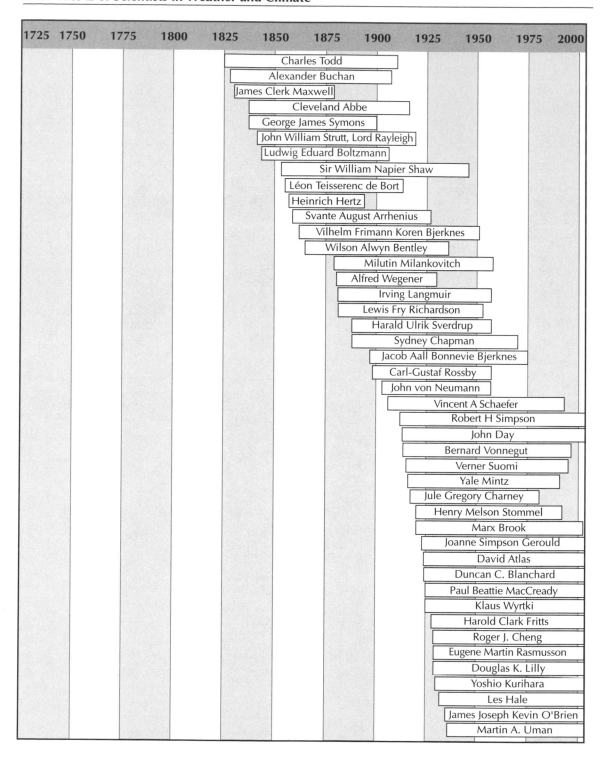

1725	1750	1775	1800	1825	1850	1875	1900	1925	1950	1975	2000

Charles Todd

Alexander Buchan

James Clerk Maxwell

Cleveland Abbe

George James Symons

John William Strutt, Lord Rayleigh

Ludwig Eduard Boltzmann

Sir William Napier Shaw

Léon Teisserenc de Bort

Heinrich Hertz

Svante August Arrhenius

Vilhelm Frimann Koren Bjerknes

Wilson Alwyn Bentley

Milutin Milankovitch

Alfred Wegener

Irving Langmuir

Lewis Fry Richardson

Harald Ulrik Sverdrup

Sydney Chapman

Jacob Aall Bonnevie Bjerknes

Carl-Gustaf Rossby

John von Neumann

Vincent A Schaefer

Robert H Simpson

John Day

Bernard Vonnegut

Verner Suomi

Yale Mintz

Jule Gregory Charney

Henry Melson Stommel

Marx Brook

Joanne Simpson Gerould

David Atlas

Duncan C. Blanchard

Paul Beattie MacCready

Klaus Wyrtki

Harold Clark Fritts

Roger J. Cheng

Eugene Martin Rasmusson

Douglas K. Lilly

Yoshio Kurihara

Les Hale

James Joseph Kevin O'Brien

Martin A. Uman

1725	1750	1775	1800	1825	1850	1875	1900	1925	1950	1975	2000

Peter Herwig Handel

Jaan Salm

Brian Alfred Tinsley

Elbert Friday, Jr.

Kristina Katsaros

John Stuart Godfrey

Michael Patrick McCormick

Michael McIntyre

Ronald L. Holle

Eugenia Enriqueta Kalnay

S. George H. Philander

Richard A. Anthes

Mark Alan Cane

Kuo-Nan Liou

Mingxing Wang

Davis Daniel Sentman

Patrick King

Margaret Anne LeMone

Rumyana Mitseva-Nikolova

Sergey Anisimov

Rafael Luis Bras

Patrick Minnis

Andrei Sinkevich

Brian Swanson

Earle Rolfe Williams

Bruce Wielicki

Philippe Richard

Kerry Emanuel

Piers John Sellers

Vladimir A. Rakov

Susan Solomon

Douglas Mach

Antony J. Surtees

David J. Stensrud

Daohong Wang

R. G. Harrison

M. Stolzenburg

Redy Mardiana

Heidi Tierney

Bard Zajac

acid rain (acid precipitation) Because pure precipitation is slightly acidic (due to the reaction between water droplets and carbon dioxide, creating carbonic acid) with a potential hydrogen (pH) of 5.6, acid precipitation refers to precipitation with a pH less than 5.6. Acid precipitation includes rain, fog, snow, and dry deposition. Anthropogenic pollutants (carbon dioxide, carbon monoxide, ozone, nitrogen and sulfur oxides, and hydrocarbons) react with water vapor to produce acid precipitation. These pollutants come primarily from burning coal and other fossil fuels. Sulfur dioxide, which reacts readily with water vapor and droplets (that is, has a short residence time in the atmosphere as a gas), has been linked to the weathering (eating away) of marble structures and the acidification of freshwater lakes (consequently killing fish). Natural interactions within the biosphere can also lead to acid precipitation.

advect A horizontal movement of a mass of fluid, such as ocean or air currents. Can also refer to the horizontal transport of something (for example, pollution, phytoplankton, ice, or even heat) by such movement. Advection is the transfer of air mass properties by the velocity field of winds.

albedo The ratio of the outgoing solar radiation reflected by an object to the incoming solar radiation incident upon it.

anticyclone A high-pressure area where winds blow clockwise in the Northern Hemisphere and counterclockwise in the Southern Hemisphere. See also CYCLONE; WIND.

atmosphere The air surrounding the Earth, described as a series of shells or layers of different characteristics. The atmosphere, composed mainly of nitrogen and oxygen with traces of carbon dioxide, water vapor, and other gases, acts as a buffer between Earth and the Sun. The layers—troposphere, stratosphere, mesosphere, thermosphere, and exosphere—vary around the globe and in response to seasonal changes.

aurora borealis (northern lights) The luminous, radiant emission from the upper atmosphere over middle and high latitudes is centered around the Earth's magnetic poles. Often seen on clear winter nights in a variety of shapes and colors.

backscatter Process by which up to 25 percent of radiant energy from the Sun is reflected or scattered away from the surface by clouds. Also radar echo that is reflected, or scattered, at 180 degrees to the direction of the incident wave.

baroclinic The state of the atmosphere where surfaces of constant pressure intersect surfaces of constant density.

baroclinic zone   A region in which a temperature gradient exists on a constant pressure surface. Baroclinic zones are favored areas for strengthening and weakening systems.

biogenic   Produced by natural processes. Usually used in the context of emissions that are produced by plants and animals.

blackbody   An ideal emitter that radiates energy at the maximum possible rate per unit area at each wavelength for any given temperature. A blackbody also absorbs all the radiant energy incident on it; that is, no energy is reflected or transmitted.

boundary layer   The layer within which the effects of friction are significant. Regarding the Earth, it is the layer considered to be about the lowest 1 or 2 kilometers of the atmosphere. Here, temperatures are most strongly affected by daytime insolation and night radiational cooling, and winds are affected by friction with the Earth's surface. The effects of friction dissipate gradually with height, so the top of this layer cannot be specifically defined. A thin layer immediately above the Earth's surface is the surface boundary layer, or simply the surface layer, and is only a part of the planetary boundary layer. It represents the layer within which friction effects are more or less constant throughout and is roughly 10 meters thick.

carbon dioxide   A minor but very important component of the atmosphere, carbon dioxide traps infrared radiation. Atmospheric CO_2 has increased about 25 percent since the early 1800s, with an estimated increase of 10 percent since 1958 (burning fossil fuels is the leading cause of increased CO_2, deforestation the second major cause). The increased amounts of CO_2 in the atmosphere enhance the greenhouse effect, blocking heat from escaping into space and contributing to the warming of Earth's lower atmosphere.

charge separation   The phenomenon within clouds that cause lightning to occur.

climate change   The term *climate change* is sometimes used to refer to all forms of climatic inconsistency, but because the Earth's climate is never static, the term is more properly used to imply a significant change from one climatic condition to another. In some cases, *climate change* has been used synonymously with the term *global warming*; scientists, however, tend to use the term in the wider sense to include also natural changes in climate.

cloud albedo   Reflectivity that varies from less than 10 percent to more than 90 percent of the insolation and depends on drop sizes, liquid water content, water-vapor content, thickness of the cloud, and the Sun's zenith angle. The smaller the drops and the greater the liquid water content, the greater the cloud albedo, if all other factors are the same.

cloud feedback   The coupling between cloudiness and surface air temperature in which a change in surface temperature could lead to a change in clouds, which could then amplify or diminish the initial temperature perturbation. For example, an increase in surface air temperature could increase the evaporation; this, in turn, might increase the extent of cloud cover. Increased cloud cover would reduce the solar radiation reaching the Earth's surface, thereby lowering the surface temperature. This is an example of negative feedback and does not include the effects of long-wave radiation or the advection in the oceans and the atmosphere, which must also be considered in the overall relationship of the climate system.

cloud forcing   The difference between the radiation budget components for average cloud conditions and cloud-free conditions. Roughly speaking, clouds increase the albedo from 15 to 30 percent, which results in a reduction of absorbed solar radiation of about 50 W/m². This cooling is offset somewhat by the greenhouse effect of clouds, which reduces the OLR by about 30 W/m^2, so the net cloud forcing of the radiation budget is a loss of about 20 W/m^2. Were the clouds to be removed with all else remaining the same, the Earth would gain this last amount in net radiation and begin to warm up.

clouds   A visible cluster of tiny water and/or ice particles in the atmosphere. Clouds may be classified on their visible appearance, height, or form.

cloud seeding   The introduction of artificial substances (usually dry ice or silver iodide) into a cloud for the purpose of either modifying its development or increasing its precipitation. First developed by Vincent Schaefer (dry ice) and Bernard Vonnegut (silver iodide).

condensation nuclei   Tiny particles in which water vapor condenses on their surfaces in the atmosphere.

contrails   Condensation trails. Artificial clouds made by the exhaust of jet aircraft.

convection   The rising of warm air and the sinking of cool air. Heat mixes and moves air. When a layer of air receives enough heat from the Earth's surface, it expands and moves upward. Under it flows colder, heavier air that is then warmed, expands, and rises. The warm, rising air cools as it reaches higher, cooler regions of the atmosphere and begins to sink. Convection causes local breezes, winds, and thunderstorms.

convective storms   Storms that produce a circular motion of air that result when warm air rises and then is replaced with cooler air.

convergence   Movement of the wind resulting in a horizontal influx of air in a specific region. Convergent winds in lower levels are associated with upper movement. The opposite of divergence.

(Oceanography) When waters of different origins come together at a point or along a line (convergence line), the denser water from one side sinks under the lighter water from the other side. The ocean convergence lines are the polar, subtropical, tropical, and equatorial. See also DIVERGENCE.

coriolis force   The apparent tendency of a freely moving particle to swing to one side when its motion is referred to a set of axes that is itself rotating in space, such as Earth. The acceleration is perpendicular to the direction of the speed of the article relative to the Earth's surface and is directed to the right in the Northern Hemisphere. Winds are affected by rotation of the Earth so that instead of a wind blowing in the direction it starts, it turns to the right of that direction in the Northern Hemisphere; left in the Southern Hemisphere.

coupled system   Two or more processes that affect one another.

cryosphere   One of the interrelated components of the Earth's system, the cryosphere is frozen water in the form of snow, permanently frozen ground (permafrost), floating ice, and glaciers. Fluctuations in the volume of the cryosphere cause changes in ocean sea level, which directly impact the atmosphere and biosphere.

cyclone   An area of low pressure where winds blow counterclockwise in the Northern Hemisphere and clockwise in the Southern Hemisphere. See also ANTICYCLONE; WIND.

diel   Varying on a day/night basis.

divergence   Movement of the wind that results in a horizontal air current coming from a particular region.

Doppler radar   The weather radar system that uses the Doppler shift of radio waves to detect air motion that can result in tornadoes and precipitation, as previously developed weather radar systems do. It can also measure the speed and direction of rain and ice, as well as detect the formation of tornadoes sooner than older radars.

downwelling   The process of accumulation and sinking of warm surface waters along a coastline. A change of air flow of the atmosphere can result in the sinking or downwelling of warm surface water. The resulting reduced nutrient supply near the surface affects the ocean productivity and meteorological conditions of the coastal regions in the downwelling area.

eccentricity   One of six Keplerian elements, it describes the shape of an orbit. In the Keplerian orbit model, the satellite orbit is an ellipse, with eccentricity defining the "shape" of the ellipse. When e = 0, the ellipse is a circle. When e is very near 1, the ellipse is very long and skinny.

El Niño   A warming of the surface waters of the eastern equatorial Pacific that occurs at irregular intervals of two to seven years, usually lasting one to two years. Along the west coast of South America, southerly winds promote the upwelling of cold, nutrient-rich water that sustains large fish populations that sustain abundant sea birds, whose droppings support the fertilizer industry. Near the end of each calendar year, a warm current of nutrient-pool tropical water replaces the cold, nutrient-rich surface water. Because this condition often occurs around Christmas, it was named El Niño (Spanish for "boy child," referring to the Christ child). In most years, the warming lasts only a few weeks or a month, after which the weather patterns return to normal and fishing improves. However, when El Niño conditions last for many months, more-extensive ocean warming occurs and economic results can be disastrous. El Niño has been linked to wetter, colder winters in the United States, drier, hotter summers in South America and Europe, and drought in Africa.

electromagnetic radiation (EM)   Energy propagated as time-varying electric and magnetic fields. These two fields are inextricably linked as a single entity because time-varying electric fields produce time-varying magnetic fields and vice versa. Light and radar are examples of electromagnetic radiation differing only in their wavelengths (or frequency). Electric and magnetic fields propagate through space at the speed of light.

electromagnetic spectrum   The entire range of radiant energies or wave frequencies from the longest to the shortest wavelengths—the categorization of solar radiation. Satellite sensors collect this energy, but what the detectors capture is only a small portion of the entire electromagnetic spectrum. The spectrum usually is divided into seven sections: radio, microwave, infrared, visible, ultraviolet, X-ray, and gamma-ray radiation.

electromagnetic wave   Method of travel for radiant energy (all energy is both particles and waves), so called because radiant energy has both magnetic and electrical properties. Electromagnetic waves are produced when electric charges change their motion. Whether the frequency is high or low, all electromagnetic waves travel at 300 million meters per second.

electron avalanche   The phenomenon of continuous production of electrons—throughout the environment in the air itself. Also known as electron cascade.

elves   Bright, short lightning flashes that appear for only a thousandth of a second and appear above the clouds at the edge of space.

emissivity   The ratio of the radiation emitted by a surface to that emitted by a black body at the same temperature.

energy balance models   (EBM) An analytical technique to study the solar radiation incident on the Earth in which explicit calculations of atmospheric motions are omitted. In the zero-dimensional models, only the incoming and outgoing radiation is considered. The outgoing infrared radiation is a linear function of global mean surface air temperature, and the reflected solar radiation is dependent on the surface albedo. The albedo is a step function of the global mean surface air temperatures, and equilibrium temperatures are computed for a range of values of the solar constant. The one-dimensional models have surface air temperature as a function of latitude. At each latitude, a balance between incoming and outgoing radiation and horizontal transport of heat is computed.

energy budget A quantitative description of the energy exchange for a physical or ecological system. The budget includes terms for radiation, conduction, convection, latent heat, and sources and sinks of energy.

enhanced greenhouse effect The natural greenhouse effect has been enhanced by anthropogenic emissions of greenhouse gases. Increased concentrations of carbon dioxide, methane, and nitrous oxide, CFCs, HFCs, PFCs, SF6, NF3, and other photochemically important gases caused by human activities such as fossil-fuel consumption and adding waste to landfills, trap more infrared radiation, thereby exerting a warming influence on the climate. See also CLIMATE CHANGE; GLOBAL WARMING.

ENSO (El Niño–Southern Oscillation) Interacting parts of a single global system of climate fluctuations. ENSO is the most prominently known source of interannual variability in weather and climate around the world, though not all areas are affected. The Southern Oscillation (SO) is a global-scale seesaw in atmospheric pressure between Indonesia/North Australia and the southeast Pacific. In major warm events, El Niño warming extends over much of the tropical Pacific and

becomes clearly linked to the SO pattern. Many of the countries most affected by ENSO events are developing countries with economies that are largely dependent upon their agricultural and fishery sectors as a major source of food supply, employment, and foreign exchange. New capabilities to predict the onset of ENSO event can have a global impact. Although ENSO is a natural part of the Earth's climate, whether its intensity or frequency may change as a result of global warming is an important concern.

exosphere The uppermost layer of the atmosphere, its lower boundary is estimated at 500 kilometers to 1000 kilometers above the Earth's surface. It is only from the exosphere that atmospheric gases can, to any appreciable extent, escape into outer space.

external forcing Influence on the Earth system (or one of its components) by an external agent, such as solar radiation, or the impact of extraterrestrial bodies, such as meteorites.

fair-weather electric current The fair-weather electric current is equivalent to a leaking current occurring between two spherical plates; that is, the space charge moves downward through the atmosphere (dielectric) between the ionosphere and Earth (plates). This net downward transport of positive charge occurs in a current between 1,350 to 1,800 amperes over the entire Earth's surface. Those studying atmospheric electricity hypothesize that lightning causes this fair-weather electric current by bringing negative charge to ground (keeping the spherical capacitor charged). If lightning did not occur continuously, the Earth-ionosphere capacitor would lose its current in less than 10 minutes.

fair-weather electric field The fair-weather electric field is measured to be an average of 120 volts per meter directed downward to the ground. It decreases exponentially with height with the air density (because the dielectric strength decreases causing the atmosphere's conductivity to increase).

finite difference model (FDM) A method of modeling physical situations by breaking up one large problem into many smaller (and easier) problems.

front A boundary between two different air masses. The difference between two air masses sometimes is unnoticeable, but when the colliding air masses have very different temperatures and amounts of water in them, turbulent weather can erupt.

A cold front occurs when a cold air mass moves into an area occupied by a warmer air mass. Moving at an average speed of about 20 miles per hour, the heavier cold air moves in a wedge shape along the ground. Cold fronts bring lower temperatures and can create narrow bands of violent thunderstorms. In North America, cold fronts form on the eastern edges of high-pressure systems.

A warm front occurs when a warm air mass moves into an area occupied by a colder air mass. The warm air is lighter, so it flows up the slope of the cold air below it. Warm fronts usually form on the eastern sides of low-pressure systems, create wide areas of clouds and rain, and move at an average speed of 15 miles per hour.

When a cold front follows and then overtakes a warm front (warm fronts move more slowly than cold fronts) lifting the warm air off the ground, an occluded front forms.

General Circulation Model (GCM) A global, three-dimensional computer model of the climate system that can be used to simulate human-induced climate change. GCMs are highly complex, and they represent the effects of such factors as reflective and absorptive properties of atmospheric water vapor, greenhouse gas concentrations, clouds, annual and daily solar heating,

ocean temperatures, and ice boundaries. The most recent GCMs include global representations of the atmosphere, oceans, and land surface.

Geographic Information System (GIS) A system for archiving, retrieving, and manipulating data that has been stored and indexed according to the geographic coordinates of its elements. The system generally can utilize a variety of data types, such as imagery, maps, and tables.

geomagnetism The magnetism of the Earth.

geosphere The physical elements of the Earth's surface crust and interior.

geostrophic wind Theoretically horizontal wind blowing in a straight path, and parallel to the isobars or contours, at a constant speed. Caused when the coriolis force balances the horizontal-pressure gradient force exactly.

global climate change The long-term fluctuations in temperature, precipitation, wind, and all other aspects of the Earth's climate. External processes, such as solar-irradiance variations, variations of the Earth's orbital parameters (eccentricity, precession, and inclination), lithosphere motions, and volcanic activity, are factors in climatic variation. Internal variations of the climate system also produce fluctuations of sufficient magnitude and variability to explain observed climate change through the feedback processes interrelating the components of the climate system.

global electric circuit A massive electrical circuit that literally covers the globe with a voltage difference between ground and ionosphere of 200,000 to 500,000 volts (200 to 500 kV).

global positioning system (GPS) A system consisting of 25 satellites in six orbital planes at 20,000 kilometers of altitude with 12-hour periods, used to provide highly precise position, velocity, and time information to users anywhere on Earth or in its neighborhood at any time.

global variables   Functions of space and time that describe the large-scale state and evolution of the Earth system. The Earth system's geosphere, hydrosphere, atmosphere, and biosphere and their components are, or potentially are, global variables.

global warming   An increase in the near surface temperature of the Earth. Global warming has occurred in the distant past as the result of natural influences, but the term is most often used to refer to the warming predicted to occur as a result of increased emissions of greenhouse gases. Scientists generally agree that the Earth's surface has warmed by about 1 degree Fahrenheit in the past 140 years. The Intergovernmental Panel on Climate Change (IPCC) recently concluded that increased concentrations of greenhouse gases are causing an increase in the Earth's surface temperature and that increased concentrations of sulfate aerosols have led to relative cooling in some regions, generally over and downwind of heavily industrialized areas. See also CLIMATE CHANGE; ENHANCED GREENHOUSE EFFECT.

graupel   Snow pellet. A small white ice particle that falls as precipitation and breaks apart easily when it lands on a surface. Ice particles between 2 and 5 millimeters in diameter that form in a cloud often by the process of accretion. Snowflakes that become rounded pellets due to riming are called graupel or snow pellets.

greenhouse effect   The warming of an atmosphere by its absorbing and remitting infrared radiation while allowing short-wave radiation to pass on through.

Certain gaseous components of the atmosphere, called greenhouse gases, transmit the visible portion of solar radiation but absorb specific spectral bands of thermal radiation emitted by the Earth. The theory is that terrain absorbs radiation, heats up, and emits longer wavelength thermal radiation that is prevented from escaping into space by the blanket of carbon dioxide and other greenhouse gases in the atmosphere. As a result, the climate warms. Because atmospheric and oceanic circulations play a central role in the climate of the Earth, improving our knowledge about their interaction becomes essential.

greenhouse gas   A gaseous component of the atmosphere contributing to the greenhouse effect. Greenhouse gases are transparent to certain wavelengths of the Sun's radiant energy, allowing them to penetrate deep into the atmosphere or all the way into the Earth's surface. Greenhouse gases and clouds prevent some of infrared radiation from escaping, trapping the heat near the Earth's surface where it warms the lower atmosphere. Alteration of this natural barrier of atmospheric gases can raise or lower the mean global temperature of the Earth.

Greenhouse gases include carbon dioxide, methane, nitrous oxide, chlorofluorocarbons, and water vapor. Carbon dioxide, methane, and nitrous oxide have significant natural and human sources, but only industries produce chlorofluorocarbons. Water vapor has the largest greenhouse effect, but its concentration in the troposphere is determined within the climate system. Water vapor will increase in response to global warming, which in turn may further enhance global warming.

gulf stream   A warm, swift ocean current that flows along the coast of the Eastern United States and makes Ireland, Great Britain, and the Scandinavian countries warmer than they would be otherwise.

heat balance   The equilibrium existing between the radiation received and emitted by a planetary system.

heat island effect   A dome of elevated temperatures over an urban area, caused by the heat absorbed by structures and pavement.

hydrochlorofluorocarbon (HCFC)   One of a class of compounds used primarily as a CFC substitute. Work on CFC alternatives began in the late 1970s after the first warnings of CFC damage to stratospheric ozone. By adding hydrogen to the chemical formulation, chemists made CFCs less stable in the lower atmosphere, enabling the CFCs to break down before reaching the ozone layer. However, HCFCs do release chlorine and have contributed more to atmospheric chlorine buildup than originally predicted. Development of non–chlorine-based chemical compounds as a substitute for CFCs and HCFCs continues.

hydrologic cycle   The process of evaporation, vertical and horizontal transport of vapor, condensation, precipitation, and the flow of water from continents to oceans. It is a major factor in determining climate through its influence on surface vegetation, the clouds, snow and ice, and soil moisture. The hydrologic cycle is responsible for 25 to 30 percent of the midlatitudes' heat transport from the equatorial to polar regions.

hydrology   The science that deals with global water (both liquid and solid), its properties, circulation, and distribution, on and under the Earth's surface and in the atmosphere, through evapotranspiration or is discharged into oceans.

hydrosphere   The totality of water encompassing the Earth, comprising all the bodies of water, ice, and water vapor in the atmosphere.

idempotent   Acting as if used only once, even if used multiple times.

infrared radiation (IR)   Infrared is electromagnetic radiation whose wavelength spans the region from about 0.7 to 1000 micrometers (longer than visible radiation, shorter than microwave radiation). Remote sensing instruments work by sensing radiation that is naturally emitted or reflected by the Earth's surface or from the atmosphere or by sensing signals transmitted from a satellite and reflected back to it. In the visible and near-infrared regions, surface chemical composition, vegetation cover, and biological properties of surface matter can be measured. In the midinfrared region, geological formations can be detected because of the absorption properties related to the structure of silicates. In the far infrared, emissions from the Earth's atmosphere and surface offer information about atmospheric and surface temperatures, water vapor, and other trace constituents in the atmosphere. Because IR data are based on temperatures rather than visible radiation, the data may be obtained day or night.

insolation   Solar radiation incident upon a unit horizontal surface on or above the Earth's surface.

ionization   The process where a neutral or uncharged atom or molecule (of gases, liquids, solids) acquires a charge, becoming an ion.

isobars   Lines drawn on a weather map joining places of equal barometric pressure.

isothermal   Of or indicating equality of temperature.

isotherms   Lines connecting points of equal temperature on a weather map.

jet stream   Rivers of high-speed air in the atmosphere. Jet streams form along the boundaries of global air masses where there is a significant difference in atmospheric temperature. The jet streams may be several hundred miles across and 1–2 miles deep at an altitude of 8–12 miles. They generally move west to east and are strongest in the winter with core wind speeds as high as 250 miles per hour. Changes in the jet stream indicate changes in the motion of the atmosphere and weather.

jets   fast-moving sprays of blue light that explode upward from the top of storm clouds to a height of about 20 miles.

La Niña   A period of stronger-than-normal trade winds and unusually low sea-surface temperatures in the central and eastern tropical Pacific Ocean; the opposite of El Niño.

latent heat   The heat that is either released or absorbed by a unit mass of a substance when it undergoes a change of state, such as during evaporation, condensation, or sublimation.

lidar   Acronym for Light Detection and Ranging, a technique for performing accurate remote measurements of atmospheric trace-gas concentration over ranges of several meters to tens of kilometers. This is done by probing the absorption lines of the gases with narrow spectral laser radiation, using the differential absorption lidar technique.

light
1. Form of radiant energy that acts upon the retina of the eye, optic nerve, and so on, making sight possible. This energy is transmitted at a velocity of about 186,000 miles per second by wavelike or vibrational motion.

2. A form of radiant energy similar to this but not acting on the normal retina, such as ultraviolet and infrared radiation.

Interplay between light rays and the atmosphere causes us to see the sky as blue and can result in such phenomena as glows, halos, arcs, flashes, and streamers.

lightning   A natural electric discharge that occurs most commonly in the atmosphere, owing to the charge separation within cumulonimbus clouds. The lightning discharge is a luminous short circuit between two locations: cloud-to-ground, intracloud, intercloud, or cloud-to-air. It is visible to humans as a thin, branching channel of light (literally a long spark) and may appear to flicker if there is more than one stroke of current. A cloud-to-ground discharge is of most interest because of its power; it can kill people, start forest fires, cause property damage, and knock out electrical power. Temperatures in a lightning discharge can reach some 28,000°C (50,000°F), the highest, naturally occurring temperature on Earth. Lightning produces electromagnetic radiation—from ultraviolet to visible to infrared to radio waves.

lithosphere   The component of the Earth's surface comprising the rock, the soil, and sediments. It is a relatively passive component of the climate system, and its physical characteristics are treated as fixed elements in the determination of climate.

long-wave radiation   The radiation emitted in the spectral wavelength greater than 4 micrometers, corresponding to the radiation emitted from the Earth and atmosphere. It is sometimes referred to as terrestrial radiation or infrared radiation, although somewhat imprecisely.

low or low-pressure system   A horizontal area where the atmospheric pressure is less than it is in adjacent areas. Because air always moves from areas of high pressure to areas of low pressure, air from these adjacent areas of higher pressure will move toward the low pressure area to equalize the pressure. This inflow of air toward the low will be affected by the Earth's rotation and will cause the air to spiral inward in a counterclockwise direction in the Northern Hemisphere. The air eventually rises near the center of the low, causing cloudiness and precipitation. See also CORIOLIS FORCE.

The air in a low rotates in a counterclockwise direction in the Northern Hemisphere and in a clockwise direction in the Southern Hemisphere. Low-pressure cells are called cyclones.

magnetosphere The region surrounding a celestial body where its magnetic field controls the motions of charged particles. The Earth's magnetic field is dipolar in nature; that is, it behaves as if produced by a giant bar magnet located near the center of the planet with its north pole tilted several degrees from Earth's geographic north pole.

   The Earth's magnetic field presents an obstacle to the solar wind, as a rock in a running stream of water. This obstacle is called a bow shock. The bow shock slows down, heats, and compresses the solar wind, which then flows around the rest of Earth's magnetic field.

mesoscale Scale of meteorological phenomena that vary in size up to 100 kilometers.

mesoscale meteorology The study of weather features, possessing the horizontal dimensions from 2 to 2,000 kilometers (1.2 to 1,250 miles) and the time scale from 1 hour to several days.

mesopause The upper boundary of the mesosphere where the temperature of the atmosphere reaches its lowest point.

mesosphere The atmospheric layer above the stratosphere, extending from about 50 to 85 kilometers in altitude. The temperature generally decreases with altitude.

micrometeorology The study of weather features, possessing the horizontal dimensions from 2 to 2,000 meters (6.56 to 6,560 feet) and the time scale from 1 to 3,600 seconds. Includes such weather phenomena as dust devils, tornadoes, short gravity waves, thermals, wakes, and plumes and is considered the study of turbulence, diffusion, and heat transfer within the atmosphere.

middle infrared Electromagnetic radiation between the near infrared and the thermal infrared, about 2–5 micrometers.

morphology The form and structure of an organism or object or one of its parts.

near infrared Electromagnetic radiation with wavelengths from just longer than the visible (about 0.7 micrometers) to about 2 micrometers. See also ELECTROMAGNETIC SPECTRUM.

nitrogen oxides (NOx) Gases consisting of one molecule of nitrogen and varying numbers of oxygen molecules. Nitrogen oxides are produced in the emissions of vehicle exhausts and from power stations. In the atmosphere, nitrogen oxides can contribute to formation of photochemical ozone (smog), can impair visibility, and can have health consequences; they are thus considered pollutants.

nitrous oxide (N_2O) A powerful greenhouse gas with a global warming potential of 320. Major sources of nitrous oxide include soil cultivation practices, especially the use of commercial and organic fertilizers; fossil fuel combustion; nitric acid production; and biomass burning.

North Atlantic Oscillation (NAO) A large seesaw in atmospheric mass between the subtropical high located near the Azores and the subpolar low near Iceland.

nowcasting Short-term weather forecasts varying from minutes to a few hours.

nucleation The coalescence of molecules or small particles that is the beginning of a crystal or a similar small deposit.

occluded front (occlusion) A composite of two fronts formed as a cold front overtakes a warm front. A cold occlusion results when the coldest air is behind the cold front. The cold front undercuts the warm front and, at the Earth's surface, the coldest air replaces less-cold air.

A warm occlusion occurs when the coldest air lies ahead of the warm front. Because the cold front cannot lift the colder air mass, it rides piggyback up on the warm front over the coldest air.

optical thickness (optical depth)   In calculating the transfer of radiant energy, the mass of an absorbing or emitting material lying in a vertical column of unit cross-sectional area and extending between two specified levels. Also, the degree to which a cloud prevents light from passing through it; the optical thickness then depends on the physical constitution (crystals, drops, and/or droplets), the form, the concentration, and the vertical extent of the cloud.

ozone   Molecule made of three oxygen atoms. Most ozone in Earth's atmosphere is in the ozone layer of the stratosphere, where it is generated by short-wave solar radiation after the dissociation of molecular oxygen. Ozone in the troposphere is also generated by humans in photochemical reactions associated with urban air pollution. This tropospheric ozone is an irritant to the eyes and the throat and is detrimental to plant surfaces.

ozone hole   A large area of intense stratospheric ozone depletion over the Antarctic continent that typically occurs annually between late August and early October and generally ends in mid-November. This severe ozone thinning has increased conspicuously since the late 70s and early 80s. This phenomenon is the result of chemical mechanisms initiated by humanmade chlorofluorocarbons (See also CFCs). Continued buildup of CFCs is expected to lead to additional ozone loss worldwide.

The thinning is focused in the Antarctic because of particular meteorological conditions there. During Austral spring (September and October in the Southern Hemisphere), a belt of stratospheric winds encircles Antarctica, essentially isolating the cold stratospheric air there from the warmer air of the middle latitudes. The frigid air permits the formation of ice clouds that facilitate chemical interactions among nitrogen, hydrogen, and chlorine (elevated from CFCs) atoms, the end product of which is the destruction of ozone.

ozone layer   The layer of ozone that begins approximately 15 kilometers above Earth and thins to an almost negligible amount at about 50 kilometers, shields the Earth from harmful ultraviolet radiation from the Sun. The highest natural concentration of ozone (approximately 10 parts per million by volume) occurs in the stratosphere at approximately 25 kilometers above Earth. The stratospheric ozone concentration changes throughout the year as stratospheric circulation changes with the seasons. Natural events such as volcanoes and solar flares can produce changes in ozone concentration, but humanmade changes are of the greatest concern.

ozone minihole(s)   Rapid, transient, polar-ozone depletion. These depletions, which take place over a 50-kilometer squared area, are caused by weather patterns in the upper troposphere. The decrease in ozone during a minihole event is caused by transport, with no chemical depletion of ozone. However, the cold stratospheric temperatures associated with weather systems can cause clouds to form that can lead to the conversion of chlorine compound from inert to reactive forms. These chlorine compounds can then produce longer-term ozone reductions after the minihole has passed.

particulates   Very small pieces of solid or liquid matter such as particles of soot, dust, fumes, mists or aerosols. The physical characteristics of particles and how they combine with other particles are part of the feedback mechanisms of the atmosphere.

photochemistry   The study of the impact of light on certain chemical molecules.

photodissociation   A chemical reaction involving sunlight in which molecules are split into their constituent atoms. Also known as *photolysis*.

photon   A quantum (smallest unit in which waves may be emitted or absorbed) of light.

photovoltaic   Capable of producing a voltage when exposed to radiant energy, especially light.

planetary albedo   The fraction of incident solar radiation that is reflected by a planet and returned to space. The planetary albedo of the Earth-atmosphere system is approximately 30 percent, most of which is due to backscatter from clouds in the atmosphere.

planetary boundary layer   The turbulent layer of atmosphere occupying the lowest few hundred meters of the atmosphere.

planetary wave   A large wave in the atmospheric circulation, the polar jet stream, and the westerly winds, characterized by a great length and a significant amplitude that extends from the middle to the upper troposphere. Also known as long wave or Rossby wave.

pooling   A thin, surface-based layer of relatively cool air, generated by diabatic processes.

positive cloud to ground   A cloud-to-ground lightning flash that delivers positive charge to the ground, as opposed to the more-common negative charge. These appear more frequently in some severe thunderstorms.

positive polarity lightning   When cloud-to-ground lightning is predominantly of positive polarity, the lightning flash discharges lower positive charge to ground.

precession   The comparatively slow torquing of the orbital planes of all satellites with respect to the Earth's axis, due to the bulge of the Earth at the equator that distorts the Earth's gravitational field. Precession is manifest by the slow rotation of the line of nodes of the orbit (westward for inclinations less than 90 degrees and eastward for inclinations greater than 90 degrees).

precipitation   The liquid or solid form of water that falls and reaches ground underneath clouds. Examples of precipitation are rain (light, moderate, heavy), drizzle, snow, hail, ice pellets, and graupel.

pulse   A short burst of radar (or sonar), typically measured as a function of time, distance, or power.

pyrogenic   Resulting from fire activities. Usually used in the context of emissions that are produced by fires, for example, smoke from fires.

radiation balance   The difference between the absorbed solar radiation and the net infrared radiation.

radiation budget   A measure of all the inputs and outputs of radiative energy relative to a system, such as Earth.

radiant energy (radiation)   Energy propagated in the form of electromagnetic waves that do not need molecules to propagate them; in a vacuum, they travel at nearly 300,000 kilometers per second.

radiative cooling   Cooling process of the Earth's surface and adjacent air that occurs when infrared (heat) energy radiates from the surface of the Earth upward through the atmosphere into space. Air near the surface transfers its thermal energy to the nearby ground through conduction so that radiative cooling lowers the temperature of both the surface and the lowest part of the atmosphere.

radiative forcing A change in the balance between incoming solar radiation and outgoing infrared radiation. Without any radiative forcing, solar radiation coming to the Earth would continue to be approximately equal to the infrared radiation emitted from the Earth. The addition of greenhouse gases traps and increased fraction of the infrared radiation, reradiating it back toward the surface and creating a warming influence (that is, positive radiative forcing because incoming solar radiation will exceed outgoing infrared radiation).

radiative transfer Theory dealing with the propagation of electromagnetic radiation through a medium.

remote sensing The application of distant devices to infer local properties, usually through electromagnetic radiation. For example, instruments on board Earth-orbiting satellites determine atmospheric properties such as temperature, cloudiness, and moisture content.

scattering The process by which electromagnetic radiation interacts with and is redirected by the molecules of the atmosphere, ocean, or land surface. The term is frequently applied to the interaction of the atmosphere on sunlight, which causes the sky to appear blue (because light near the blue end of the spectrum is scattered much more than light near the red end).

scatterometer A high-frequency radar instrument that transmits pulses of energy toward the ocean and measures the backscatter from the ocean surface. It detects wind speed and direction over the oceans by analyzing the backscatter from the small wind-induced ripples on the surface of the water.

secondary electrons An electron emitted by ionization of a surface atom when a high-energy electron beam impinges a surface. Of low energy and subject to electromagnetic fields.

short-wave radiation The radiation received from the Sun and emitted in the spectral wavelengths of less than 4 microns. It is also called solar radiation.

solar radiation Energy received from the Sun is solar radiation. The energy comes in many forms, such as visible light (that which we can see with our eyes). Other forms of radiation include radio waves, heat (infrared), ultraviolet waves, and X rays. These forms are categorized within the electromagnetic spectrum.

solar wind A continuous plasma stream expanding into interplanetary space from the Sun's corona. The solar wind is present continuously in interplanetary space. After escaping from the gravitational field of the Sun, this gas flows outward at a typical speed of 400 kilometers per second to distances that are known to be beyond the orbit of Pluto. Besides affecting Earth's weather, solar activity gives rise to a dramatic visual phenomena in our atmosphere. The streams of charged particles from the Sun interact the Earth's magnetic field like a generator to create current systems with electric potentials of as much as 100,000 volts. Charged electrons are energized by this process, sent along the magnetic field lines toward Earth's upper atmosphere, excite the gases present in the upper atmosphere, and cause them to emit light, which we call the auroras. The auroras are the northern (aurora borealis) and southern (aurora Australis) lights.

sounding rocket   A sounding rocket takes a parabolic path into space but does not orbit the earth. It is a suborbital rocket.

southern oscillation   A large-scale atmospheric and hydrospheric fluctuation centered in the equatorial Pacific Ocean. It exhibits a nearly annual pressure anomaly, alternatively high over the Indian Ocean and high over the South Pacific. Its period is slightly variable, averaging 2.33 years. The variation in pressure is accompanied by variations in wind strengths, ocean currents, sea-surface temperatures, and precipitation in the surrounding areas. El Niño and La Niña occurrences are associated with the phenomenon.

spectrometer   A spectroscope that is equipped with scales for measuring wavelengths or indexes of refraction.

spectrophotometer   A device for measuring the relative amounts of radiant energy or radiant flux as a function of wavelength.

spectrum   Any of various arrangements of colored bands or lines, together with invisible components at both ends of the spectrum, similarly formed by light from incandescent gases or other sources of radiant energy, which can be studied by a spectrograph.

sprite   A red light flash that appears as high as 60 miles above thunderstorms.

squall line   A solid or nearly solid line or band of active thunderstorms.

stratosphere   Region of the atmosphere between the troposphere and mesosphere, having a lower boundary of approximately 8 kilometers at the poles to 15 kilometers at the equator and an upper boundary of approximately 50 kilometers. Depending on latitude and season, the tempera-ture in the lower stratosphere can increase, be isothermal, or even decrease with altitude, but the temperature in the upper stratosphere generally increases with height due to absorption of solar radiation by ozone.

sublimation   The change of state from ice to water vapor or water vapor to ice.

subsidence   In weather-forecasting terminology, refers to sinking motions of air masses. It could also refer to sinking motions within fluids or bodies of water.

surface air temperature   The temperature of the air near the surface of the Earth, usually determined by a thermometer in an instrument shelter about 2 meters above the ground. The true daily mean, obtained from a thermograph, is approximated by the mean of 24 hourly readings and may differ by 1.0 degrees C from the average, based on minimum and maximum readings. The global average surface air temperature is 15 degrees C.

synoptic chart   Chart showing meteorological conditions over a region at a given time; weather map.

synoptic scale   Size of the migratory systems of high or low pressure in the lowest troposphere, in an area of hundreds of kilometers or more.

synoptic view   The ability to see large areas at the same time.

terrestrial radiation   The total infrared radiation emitted by the Earth and its atmosphere in the temperature range of approximately 200–300K. Because the Earth is nearly a perfect radiator, the radiation from its surface varies as the fourth power of the surface's absolute temperature. Terrestrial radiation provides a major part of the potential energy changes necessary to drive the atmospheric wind system and is responsible for maintaining the surface air temperature within limits for livability.

thermocline A transition layer of water in the ocean, with a steeper vertical temperature gradient than that found in the layers of ocean above and below. The permanent thermocline separates the warm, mixed surface layer of the ocean from the cold, deep ocean water and is found between 100- and 1000-meter depths. The thermocline first appears at the 55°–60° N and S latitudes, where it forms a horizontal separation between temperate and polar waters. The thermocline reaches its maximum depth at midlatitudes and is shallowest at the equator and at its northern and southern limits. The thermocline is stably stratified, and transfer of water and carbon dioxide across this zone occurs very slowly. Thus, the thermocline acts as a barrier to the downward mixing of carbon dioxide.

thermodynamic The science of heat and temperature and of the laws governing the conversion of heat into mechanical, electrical, or chemical energy.

thermosphere The outermost shell of the atmosphere between the mesosphere and outer space; where temperatures increase steadily with altitude.

trace gas Any one of the less common gases found in the Earth's atmosphere. Nitrogen, oxygen, and argon make up more than 99 percent of the Earth's atmosphere. Other gases, such as carbon dioxide, water vapor, methane, oxides of nitrogen, ozone, and ammonia, are considered trace gases. Although relatively unimportant in terms of their absolute volume, they have significant effects on the Earth's weather and climate.

trade winds Surface air from the horse latitudes that moves back toward the equator and is deflected by the coriolis force, causing the winds to blow from the northeast in the Northern Hemisphere and from the southeast in the Southern Hemisphere. These steady winds are called trade winds because they provided trade ships with an ocean route to the New World.

tropopause The boundary between the troposphere and the stratosphere (about 8 km in polar regions and about 15 km in tropical regions), usually characterized by an abrupt change of lapse rate. The regions above the troposphere have increased atmospheric stability than those below. The tropopause marks the vertical limit of most clouds and storms.

troposphere The lower atmosphere, to a height of 8–15 km above Earth, where temperature generally decreases with altitude, clouds form, precipitation occurs, and convection currents are active. See also ATMOSPHERE.

tropospheric ozone (O_3) Ozone that is located in the troposphere and plays a significant role in the greenhouse gas effect and urban smog. See also OZONE.

trough Elongated area of low atmospheric pressure, either at the surface or in the upper atmosphere.

ultraviolet radiation The energy range just beyond the violet end of the visible spectrum. Although ultraviolet radiation constitutes only about 5 percent of the total energy emitted from the Sun, it is the major energy source for the stratosphere and mesosphere, playing a dominant role in both energy balance and chemical composition.

Most ultraviolet radiation is blocked by Earth's atmosphere, but some solar ultraviolet penetrates and aids in plant photosynthesis and helps produce vitamin D in humans. Too much ultraviolet radiation can burn the skin, cause skin cancer and cataracts, and damage vegetation.

updraft   A relatively small-scale current of air with marked upward vertical motion.

upwelling   The vertical motion of water in the ocean by which subsurface water of lower temperature and greater density moves toward the surface of the ocean. Upwelling occurs most commonly among the western coastlines of continents but may occur anywhere in the ocean. Upwelling results when winds blowing nearly parallel to a continental coastline transport the light surface water away from the coast. Subsurface water of greater density and lower temperature replaces the surface water and exerts a considerable influence on the weather of coastal regions. Carbon dioxide is transferred to the atmosphere in regions of upwelling. This is especially important in the Pacific equatorial regions, where 1–2 GtC/year may be released to the atmosphere. Upwelling also results in increased ocean productivity by transporting nutrient-rich waters to the surface layer of the ocean.

vorticity   A measure of the spin of a fluid, usually small air parcels. Absolute vorticity is the combined vorticity due to the Earth's rotation and the vorticity due to the air's circulation relative to the Earth. Relative vorticity is due to the curving of the air flow and wind shear.

water cycle   The process by which water is transpired and evaporated from the land and water, condensed in the clouds, and precipitated out onto the Earth once again to replenish the water in the bodies of water on the Earth.

water vapor   The most abundant greenhouse gas, it is the water present in the atmosphere in gaseous form. Water vapor is an important part of the natural greenhouse effect. Although humans are not significantly increasing its concentration, it contributes to the enhanced greenhouse effect because the warming influence of greenhouse gases leads to a positive water vapor feedback. In addition to its role as a natural greenhouse gas, water vapor plays an important role in regulating the temperature of the planet because clouds form when excess water vapor in the atmosphere condenses to form ice and water droplets and precipitation.

wave   The intersection of warm and cold fronts.

wave cyclone   A cyclone that forms and moves along a front, producing by its circulation a wave-like deformation of the front.

BIBLIOGRAPHY

PRINT

Abbe, Cleveland. *The Aims and Methods of Meteorological Work, Especially as Conducted by National and State Weather Services.* Baltimore: The Johns Hopkins Press, 1899.

———. *The Mechanics of the Earth's Atmosphere: A Collection of Translations.* Washington, D.C.: The Smithsonian Institution, 1910.

———. "The Meteorological Work of the U. S. Signal Service, 1870 to 1891. Report of the Chicago Meteorological Congress." 53 pp., 1893.

———. "A Short Account of the Circumstances Attending the Inception of Weather Forecast Work by the United States. April 17, 1916. Weather Bureau Topics and Personnel." Pp. 1–3, April 1916.

Abbe, Truman. *Professor Abbe and the Isobars: The Story of Cleveland Abbe, America's First Weatherman.* New York: Vantage Press, 1955.

Adams, Oscar Fay. *A Dictionary of American Authors.* 4th ed. Boston: Houghton Mifflin, 1897.

Ahrens, C. Donald. *Meteorology Today: An Introduction to Weather, Climate, and the Environment.* 5th ed. Minneapolis/St. Paul: West Pub, 1994.

———. *Essentials of Meteorology: An Invitation to the Atmosphere.* 2nd ed. Belmont, Calif.: Wadsworth, 1998.

———. *Meteorology Today: An Introduction to Weather, Climate, and the Environment.* Pacific Grove, Calif.: Brooks/Cole Pub, 2000.

———. *Workbook/Study Guide for Meteorology Today: An Introduction to Weather, Climate, and the Environment.* 6th ed. Pacific Grove, Calif.: Brooks/Cole, 2000.

Alexander of Aphrodisiacs. Eric Lewis, trans. *On Aristotle's Meteorology 4 ("Ancient Commentators on Aristotle").* Ithaca, N.Y.: Cornell University Press, 1996.

Alter, Cecil J. "National Weather Service Origins," *Bulletin of the Historical and Philosophical Society of Ohio 7,* pp. 139–85, 1949.

Amontons, G. "De la résistance causée dans les machines, Des Sciences." Académie Royale, Paris, 1699.

Arrhenius, Svante. "On the Influence of Carbonic Acid in the Air upon the Temperature of the Ground." *Philosophical Magazine, Series 5, Vol. 41, No. 251,* pp. 237–76, April 1896.

Ashford, Oliver M. *Prophet-or Professor?: The Life and Work of Lewis Fry Richardson.* Bristol and Boston: Adam Hilger, 1985.

Aspray, William. "The Mathematical Reception of the Modern Computer: John Von Neumann and the Institute for Advanced Study Computer," in Phillips, Esther R. (ed.), *Studies in the History of Mathematics, Vol. 26,* MAA, Washington, D.C., pp. 166–94, 1987.

Avery, Myron H., and Kenneth S. Boardman. "Arnold Guyot's Notes on the Geography of the Mountain District of Western North Carolina," *North Carolina Historical Review* XV, pp. 256–60.

Baker, D. James. "The Oceans 50th Anniversary." *Oceanography 5, No. 3*, pp. 154–55, 1992.

Bates, Charles C., and John F. Fuller. *America's Weather Warriors*. College Station: Texas A&M University Press, 1986.

Bathurst, Bella. *The Lighthouse Stevensons*. Canada: HarperCollins: 1999.

Bentley, W. A. "Photographing Snowflakes." *Popular Mechanics Magazine*, Vol. 37, pp. 309–12, 1922.

———. "Photomicrographa of snow crystals, and methods of reproduction." *Mon. Wea. Rev.* Vol. 46-S, pp. 359–60, August 1918.

———. "Studies of frost and ice crystals." *Mon. Wea. Rev.* Vol. 35, Nos. 8–12, pp. 348–52, 397–403, 443–44, 512–16, 584–85, August–December 1907.

———. "Studies of raindrops and raindrop phenomena." *Mon. Wea. Rev.* Vol. 32, No. 10, pp. 450–56, October 1904.

———. "Studies among the snow crystals during the winter of 1901–2, with additional data collected during previous winters." *Mon. Wea. Rev.* Vol. 30, No. 13, pp. 607–16, Annual 1902.

———. "Twenty years' study of snow crystals." *Mon. Wea. Rev.* pp. 29–35, 212–14, May 1901.

Bentley, W. A., and G. H. Perkins. "A study of snow crystals." *Appleton's Pop. Sci. Mon.* Vol. 53, No. 1, pp. 75–82, May 1898.

Birkhoff, G., et al. "Memorial Papers on John von Neumann," *Bull. AMS*, Vol. 64, No. 3, Pt. 2. 1958.

Blackmore, J., ed. *Ludwig Boltzmann: His Later Life and Philosophy, 1900–1906*. Dordrecht, Netherlands: Kluwer Academic Publisher, 1995.

Blanchard, Duncan. *The Snowflake Man: A Biography of Wilson A. Bentley*. Blacksburg, Va.: McDonald & Woodward Pub. Co., 1998.

———. "Wilson Bentley, Pioneer in Snowflake Photomicrography." *Photographic Applications in Science, Technology and Medicine*, Vol. 8, No. 3, pp. 26–28, 39–41, May 1973.

———. "The Life and Science of Alfred H. Woodcock." *Bulletin of the American Meteorological Society.* 65(5), pp. 457–463, May 1984.

———. "Wilson Bentley, the Snowflake Man." *Weatherwise*, 23(6), pp. 260–69, 1970.

Bochner, Salomon. "John von Neumann," *Biographical Memoirs*, Vol. 32, National Academy of Sciences, pp. 456–51, 1958.

Broda, E., and L. Gray. *Ludwig Boltzmann: Man, Physicist, Philosopher*. Woodbridge, Conn.: Oxbow Press, 1983.

Brown, Ralph H. "Arnold Guyot's Notes on the Southern Appalachians," *Geogr. Rev., Vol. 29*, pp. 157 f., 1939.

Brush, Stephen G., Helmut Erich Landsberg, and Martin Collins. *The History of Geophysics and Meteorology: An Annotated Bibliography*. New York: Garland Pub., 1985.

Burchfield, J. D. *Lord Kelvin and the Age of the Earth*. New York: Science History Publications, 1975.

Burlingame, Roger. *Benjamin Franklin: Envoy Extraordinary*. New York: Coward-McCann, Inc., 1967.

Chandler, T. J., and American Museum of Natural History. *The Air around Us; Man Looks at His Atmosphere*. Garden City, N.Y: Published for the American Museum of Natural History [New York] by the Natural History Press, 1969.

Clark, Ronald W. *A Biography: Benjamin Franklin*. New York: Random House, 1983.

Cohen, I. Bernard. *Benjamin Franklin: Scientist and Statesman*. New York: Charles Scribner's Sons, 1975.

Cowan, R. S. *Sir Francis Galton and the Study of Heredity in the Nineteenth Century*. New York: Garland, 1985.

Cowling, T. G. "Sydney Chapman, 1888–1970," *Biographical Memoirs of Fellows of the Royal Society*. Vol. 17: pp. 52–89, November 1971.

Cowling, T. G. "Obituary for Chapman, Sydney," *QJRAS* 11, p. 381 (1970).

Croll, James. *Climate and Time in Their Geological Relations: A Theory of Secular Changes in the Earth's Climate.* New York: Appleton and Co., 1875.

Crow, James F. "Francis Galton: count and measure, measure and count." *Genetics*, Vol. 135, pp. 1–4, 1993.

Dalton, John. *Meteorological Observations and Essays, 1793; A New System of Chemical Philosophy, 1808–10*, second volume, 1827.

Davenport, D. A. "John Dalton's First Paper and Last Experiment," *ChemMatters*, p. 14, April 1984.

Dineen, Michael P. *The Most Amazing American: Benjamin Franklin.* Waukesha, Wisconsin: Country Beautiful, 1973.

Diski, J. *Monkey's Uncle.* London: Weidenfeld & Nicolson, 1994.

Donovan, Frank R. *The Many Worlds of Benjamin Franklin.* New York: American Heritage Publishing Co., 1963.

Dunbar, G. S. "Lorin Blodget, 1923–1901." In *Geographers: Bibliographical Studies*, Vol. 5, ed. T. W. Freeman. London: Mansell, 1981.

Espy, James P. "The Philosophy of Storms." *The United States Democratic Review*, vol. 9, no. 39, September 1841.

Flamm, D. "Ludwig Boltzmann and his influence on science." *Stud. Hist. Philos. Sci.* 14 (1983), pp. 255–78.

Fleming, James Rodger. *Meteorology in America, 1800–1870.* Baltimore, Md.: John Hopkins University Press, 2000.

———. *Historical Perspectives on Climate Change.* Oxford: Oxford University Press, 1998.

Fleming, James Rodger, and The American Meteorological Society Meeting. *Historical Essays on Meteorology, 1919–1995: The Diamond Anniversary History Volume of the American Meteorological Society.* Boston, Mass.: American Meteorological Society, 1996.

Fleming, Thomas. *The Man Who Dared The Lightning: A New Look At Benjamin Franklin.* New York: William Morrow and Company, Inc., 1971.

Forrest, D. W. *Francis Galton: The Life and Work of a Victorian Genius.* London: Cambridge University Press, 1974.

Fowler, Mary J. *Great Americans.* Grand Rapids, Mich.: The Fideler Company, 1960.

Friedman, Robert Marc. *Appropriating the Weather: Vilhelm Bjerknes and Construction of a Modern Meteorology.* New York: Cornell University Press, 1989.

Friedman, Robert Marc. *The Expeditions of Harald Ulrik Sverdrup: Contexts for Shaping an Ocean Science.* La Jolla, Calif.: Scripps Institution of Oceanography UCSD, Ritter Fellowship Lecture, 1994.

Frisinger, H. Howard. *The History of Meteorology: to 1800.* New York: Science History Publications, 1977.

Galton, F. *Hereditary Genius: An Inquiry into its Laws and Consequences.* London: MacMillan. 1869.

———. *Meteorographica, or Methods of Mapping the Weather.* London: MacMillan, 1863.

Gindikin, Semen Grigorsevich. *Tales of Physicists and Mathematicians.* Boston: Birkhauser, 1988.

Gingerich, Owen. "How Galileo changed the rules of science." *Sky and Telescope*, Vol. 85, pp. 32–6, March 1993.

———. "Galileo and the phases of Venus." *Sky and Telescope*, Vol. 68, pp. 520–22, December 1984.

Glimm, J. G., and J. Impagliazzo, and I. Singer. "The Legacy of John von Neumann." Providence, R.I.: American Mathematical Society, 1990.

Goldstine, Herman H. *The Computer from Pascal to von Neumann.* Princeton, N.J.: Princeton University Press, 1972.

Grigull, U. "Fahrenheit, a Pioneer of Exact Thermometry." The Proceedings of the Eighth International Heat Transfer Conference, San Francisco, Vol. 1, pp. 9–18, 1966.

Hamer, Mick. "On the paths of genius." *New Scientist*, Vol. 141, Supp. pp. 4–7, 1994.

Havens, Barrington S. *Early history of cloud seeding; with italicized annotations by James E. Jiusto and Bernard Vonnegut.* Socorro, N.Mex.: Langmuir Laboratory, New Mexico Institute of Mining and Technology, 1978.

Hayes, Brian. "The Weatherman." *American Scientist*, Vol. 89, No. 1, pp. 10–14, 2001.

Hooykaas, R. "Science and Religion in the 17th century; Isaac Beeckman (1588–1637)," *Free University Quarterly*, Vol. 1, pp. 169–83, 1951.

Hughes, P. *A Century of Weather Service: A History of the Birth and Growth of the National Weather Service, 1870–1970.* New York: Gordon and Breach Science Publishers, Inc., 1970.

Humphreys, William J. "Origin and Growth of the Weather Service of the United States, and Cincinnati's Part Therein." *Scientific Monthly* 18, 372–82, 1924.

———. "Cincinnati's Part Therein." *Scientific Monthly*, Vol. 18, 372–82, 1924.

———. "Biographical Memoir of Cleveland Abbe, 1838–1916." *Biographical Memoirs of the National Academy of Sciences*, Vol. 8, pp. 469–508, 1919.

Iliffe, Rob. "Material doubts: Hooke, artisan culture and the exchange of information in 1670s London." *The British Journal for the History of Science*, Vol. 28, pp. 285–318, 1995.

Jaffe, Bernard, *Crucibles: The Story of Chemistry*, New York: Simon and Schuster, Inc., 1930.

Jahns, Patricia. *Joseph Henry: Father of American Electronics.* (A Ruthledge Book.) Englewood Cliffs, N.J.: Prentice Hall, 1970.

Jankovic, Vladimir. *Reading the Skies: A Cultural History of English Weather, 1650–1820.* Manchester, Eng.: Manchester University Press, 2000.

Jewell, Ralph. "The meteorological judgment of Vilhelm Bjerknes." *Social Research*, Vol. 51, pp. 783–807, August 1984.

Kevles, Daniel J. *In the Name of Eugenics.* Berkeley: University of California Press, 1985.

Kutzbach, Gisela. *The Thermal Theory of Cyclones: A History of Meteorological Thought in the Nineteenth Century, American Meteorological Society Historical Monograph Series.* Boston: American Meteorological Society, 1979.

Levin, Sheldon M. "Norwegians Led the Way in Training Wartime Weather Officers." *EOS, Transactions, American Geophysical Union*, Vol. 78, No. 52, pp. 609 et. seq., December 30, 1997.

Lewis, J. M. "Winds over the world sea: Maury and Koppen." *Bulletin of the American Meteorological Society*, Vol. 77, pp. 935–52, May 1996.

Libbey, William, Jr. "The Life and Scientific Work of Arnold Guyot." *Bulletin of the American Geographical Society* 16 (1884): 194–221.

Liljas, E., and A. H. Murphy. "Anders Ångstrom and his early papers on probability forecasting and the use/value of weather forecasts." *Bulletin of the American Meteorological Society*, Vol. 75 (7), pp. 1,227–36, 1994.

Lindzen, Richard S., Edward N. Lorenz, and George W. Platzman, eds. *The Atmosphere: A Challenge, The Science of Jule Gregory Charney.* Boston: American Meteorological Society, 1990.

Lockhart, Gary. *The Weather Companion: An Album of Meteorological History, Science, Legend, and Folklore.* New York: Wiley, 1988.

Looby, Christopher. *Benjamin Franklin.* New York: Chelsea House Publishers, 1990.

Loomis, Elias. *A Treatise on Meteorology.* New York: Harper & Brothers, 1880.

Lynch, Peter. "Richardson's Marvellous Forecast," pp. 61–73 in *The Life Cycles of Extratropical Cyclones*, M. A. Shapiro and S. Crønås, eds., Boston: American Meteorological Society, 1999.

McIntyre, D. P. *Meteorological Challenges: A History.* Ottawa: Information Canada, 1972.

McLellan, Neil J. "Robert Hooke (1635–1703): Recognizing a Sound Imagination." *The Lancet (North American Edition)*, Vol. 352, No. 9,124, pp. 312–13, July 25, 1998.

Mellersh, H. E. L. *Fitzroy of the Beagle*. New York: Mason and Lipscomb, 1968.

Meltzer, Milton. *Benjamin Franklin: The New American*. New York: Franklin Watts, 1988.

Middleton, W. E. Knowles. *Invention of Meteorological Instruments*. Baltimore: The Johns Hopkins Press, 1969.

———. *A History of the Thermometer and Its Use in Meteorology*. Baltimore: The Johns Hopkins Press, 1966.

———. *The History of the Barometer*. Baltimore: Johns Hopkins Press, 1964.

———. *A History of the Theories of Rain and Other Forms of Precipitation, Watts History of Science Library*. New York: F. Watts, 1966.

———. *Visibility in Meteorology: The Theory and Practice of the Measurement of the Visual Range*. 2nd ed. Completely Rev. and Enl. ed. Toronto: The University of Toronto Press, 1941.

Millikan, Frank. "Joseph Henry's Grand Meteorological Crusade," *Weatherwise*, Vol. 50, pp. 14–18, October/November 1997.

Mills, Eric L. "The Oceanography of the Pacific: George F. McEwen, H.U. Sverdrup and the Origin of Physical Oceanography on the West Coast of North America." *Annals of Science*, 48, No. 3, pp. 241–66, 1991.

Moore, C. B., and B. Vonnegut. "The Thundercloud," in *Lightning*, edited by R. H. Golde. New York: Academic Press, 1977.

Moore, J. T. *A History of Chemistry*. New York: McGraw-Hill, 1939.

Moran, Joseph M., and Michael D. Morgan. *Meteorology: The Atmosphere and the Science of Weather*. Upper Saddle River, N.J.: Prentice Hall, 1997.

Moyer, Albert E. *Joseph Henry: The Rise of An American Scientist*. Washington, D.C.: Smithsonian Institution Press, 1998.

Mulligan, Lotte. "Self-scrutiny and the study of nature: Robert Hooke's diary as natural history." *Journal of British Studies*, Vol. 35, pp. 311–42, July 1996.

Myhrvold, Nathan. "John Von Neumann: Computing's cold warrior." *Time*, Vol. 153, No. 12, p. 150, March 29, 1999.

Namias, Jerome. "The Early Influence of the Bergen School on Synoptic Meteorology in the United States." *Pure and Applied Geophysics*, Bergeron Memorial Volume 119, pp. 491–500, 1980/81.

Namias, Jerome. "The History of Polar Front and Air Mass Concepts in the United States: An Eyewitness Account." *Bulletin of the American Meteorological Society*, Vol. 64, No. 7, pp. 734–755, July 1983.

Nierenberg, William A. "Harald Ulrik Sverdrup, 1888–1957." *Biographical Memoirs of the National Academy of Sciences* 69 (1996): 3–38.

O'Connor, J. J., and E. F. Robertson. "Vilhelm Friman Koren Bjerknes." School of Mathematics and Statistics. University of St. Andrews, Scotland, July 2000.

Parker, G. D. "Galileo, planetary atmospheres, and prograde revolution." *Science*, Vol. 227, pp. 597–600, February 8, 1985.

Parshall, Gerald. "A calculating man." *U.S. News & World Report*, Vol. 125, No. 7, pp. 69–71, August 17–24, 1998.

Patterson, E. C. *John Dalton and the Atomic Theory*. Garden City, N.Y.: Doubleday and Co, Inc., 1970.

Pearson, Karl. *The Life, Letters, and Labours of Francis Galton*, 3 vols. New York: Cambridge University Press, 1914–30.

Philips, Norman. "Carl-Gustaf Rossby: His times, personality, and actions." *Bulletin of the American Meteorological Society*, 79, 1,097–1,112, 1998.

Phipson, Thomas Lamb. *Researches on the Past and Present History of the Earth's Atmosphere*. London: Charles Griffin & Co., 1901.

Potter, Robert R. *Benjamin Franklin.* Englewood Cliffs, N.J.: Silver Burdett Publishers, 1991.

Reingold, Nathan. "The New York State Roots of Joseph Henry's National Career." *New York History*, pp. 132–44, April 1973.

———. "Cleveland Abbe at Pulkowa: Theory and Practice in the Nineteenth Century Physical Sciences." *Archives Internationales d'Histoire des Sciences*, Vol. 17, pp. 133–47, 1964.

Revelle, Roger. "The Age of Innocence and War in Oceanography." *Oceans*, Vol. 1, pp. 6–16, 1969.

Revelle, Roger, and Walter Munk. "Harald Ulrik Sverdrup—An Appreciation." *Journal of Marine Research* Vol. 7, No. 3, pp. 127–31, 1948.

Rittner, Don, *Images of America—Albany.* Charleston, S.C.: Arcadia Press, 1999.

Romer, Robert H. "Temperature scales: Celsius, Fahrenheit, Kelvin, Réaumur, and Romer." *The Physics Teacher*, Vol. 20, pp. 450–54, October 1982.

Roscoe, H. E. *John Dalton and the Rise of Modern Chemistry.* New York: MacMillan and Co., 1895.

Rouvray, Dennis H. "John Dalton: the world's first stereochemist." *Endeavour (Oxford, England)*, Vol. 19, No. 2, pp. 52–7, 1995.

Schneider, Stephen H., ed. *Encyclopedia of Climate and Weather.* New York: Oxford University Press, 1996.

Seeger, Raymond J. *Benjamin Franklin: New World Physicist.* New York: Pergamon Press, 1973.

Seiwell, H. R. "Military Oceanography in World War II." *Military Engineer,* Vol. 39, No. 259, pp. 202–10, 1947.

Shaw, Sir Napier. *Manual of Meteorology—Vol. I, Meteorology in History.* London: Cambridge University Press, 1926.

———. *Manual of Meteorology—Vol. 2: Comparative Meteorology,* 2nd Ed. London: Cambridge University Press, 1942.

———. *Manual of Meteorology—Vol. 3: The Physical Processes of Weather.* London: Cambridge University Press, 1942.

———. *Manual of Meteorology—Vol. 4: Meteorological Calculus: Pressure and Wind.* London: Cambridge University Press, 1942.

———. "The meteorological aspects of the storm of February 26–27, 1903." *Quarterly Journal of the Royal Meteorological Society,* 29, pp. 233–58, 1903.

———. *Forecasting Weather.* London: Van Nostrand, 1911.

——— and R. G. K. Lempfert. "The life history of surface air currents." *Met. Office Memoir,* No. 174. Reprinted in *Selected Meteorological Papers of Sir Napier Shaw* (1955), MacDonald and Co., pp. 15–31, 1906.

Simonis, Doris. Editor. *Lives and Legacies: Scientists, Mathematicians, and Inventors.* New York: Oryx Press, 1999.

Simpson, J. "Global Circulation and Tropical Cloud Activity." In *The Global Role of Tropical Rainfall,* Proceedings of the International Symposium on Aqua and Planet, Vol. 129, pp. 77–92. Hampton, Va.: A. Deepak Publishing, 1992.

Smith, David C. *Climate, Agriculture, and History.* Washington, D.C.: Agricultural History Society, 1989.

Snelders, H. A. M., "Arrhenius, Svante August." In *Dictionary of Scientific Biography,* Charles C. Gillespie, ed., Vol. 1, pp. 296–302. New York: Charles Scribner's Sons, 1971.

Soter, Steven. "Galileo's Saturn." *Natural History,* Vol. 107, No. 6, p. 4, July/August 1998.

Stephens, Carlene E. *Inventing Standard Time.* Washington, D.C.: National Museum of American History, The Smithsonian Institution, 1983.

Sverdrup, Harald U. "The Pacific Ocean." *Science,* Vol. 94, No. 2,439, pp. 287–93, September 26, 1941.

———. "Vilhelm Bjerknes in Memoriam." *Tellus* Vol. 3, No. 4, pp. 217–21, November 1951.

————. "The Work at the Scripps Institution of Oceanography." *The Collecting Net* Vol. 12, No. 2, July 10, 1937.

Sverdrup, Harald U., and Martin W. Johnson, and Richard H. Fleming. *The Oceans, Their Physics, Chemistry and General Biology.* New York: Prentice Hall, 1942.

Tao, W. K. "Summary of a symposium on cloud systems, hurricanes, and TRMM: celebration of Dr. Joanne Simpson's career—the first fifty years." *Bulletin of the American Meteorological Society,* Vol. 81, No. 10, pp. 2,463–74, October 2000.

Taub, A. H., ed. *John von Neumann: Collected Works, 1903–1957,* 6 Vols., Oxford: Pergamon Press, 1961–63.

Thomson, T. *The History of Chemistry,* New York: Arno Press, 1975.

Ulam, S. "John von Neumann, 1903–1957." *Ann. Hist. Comp.,* Vol. 4, No. 2, April 1982.

Uppenbrink, Julia. "Arrhenius and global warming." *Science,* Vol. 272, pp. 1, 122, May 24, 1996.

Walz, F. J. "Present Knowledge of Meteorology and Climatology of Maryland." In *Maryland Weather Service,* Vols. I & II. Baltimore: Johns Hopkins Press, 1899.

Williams, Richard. "The tortoise and the hare: the race between evaporation and precipitation." *Weatherwise,* Vol. 52, No. 1, pp. 28–30, January/February 1999.

Williams, Frances L. *Matthew Fontaine Maury, Scientist of the Sea.* New Brunswick, N.J.: Rutgers Univ. Press, 1963.

Williams, James Thaxter. *The History of Weather.* Commack, N.Y.: Nova Science Publishers, 1999.

Wright, Esmond. *Franklin of Philadelphia.* Cambridge Mass.: Harvard University Press, 1986.

INTERNET

Academy of Technological Sciences and Engineering, *Technology in Australia 1788–1988,* Australian Science and Technology Heritage Centre, 2000. Available online. URL: http://barney.asap.unimelb.edu.au/tia/tia-dynindex.php3?EID=P000836. Posted on October 10, 2002.

Australian Science and Technology Heritage Centre, *Federation and Meteorology,* 2001. Available online. URL: http://barney.asap.unimelb.edu.au/fam/fam-dynindex.php3?EID=P000836. Posted on October 10, 2002.

Australian Science and Technology Heritage Centre, *Science and the Making of Victoria,* 2001, Available online. URL: http://barney.asap.unimelb.edu.au/smv/smv-dynindex.php3?EID=P000836.

Bellrock Lighthouse
Available online. URL: http://www.bellrock.org.uk/stevenson.html. Posted on October 10, 2002.

Benjamin Franklin: A Documentary History.
Available online. URL: http://www.english.udel.edu/lemay/franklin/. Posted on October 10, 2002.

Bentley snowflake images
Available online. URL: http://www.jericho-underhill.com/bentley.htm. Posted on October 10, 2002.

"Charles Todd—supervised construction of Overland Telegraph Line" in *The 1997 Australian Science Festival,* 1997.
Available online. URL: http://www.asap.unimelb.edu.au/bsparcs/other/asf_scientists.htm#charles. Posted on October 10, 2002.

Chew, Joe. *Storms Above The Desert. Atmospheric Research in New Mexico. 1935–1985.* Assisted by Jim Corey. Introduction by Bernard Vonnegut. University of New Mexico Press. Available online. URL: http://www.nmt.edu/~pio/StormsAbove/CoverStorms.html Posted on October 10, 2002.

Francis Galton
Available online. URL: http://www.cimm.jcu.edu.au/hist/stats/galton/. Posted on October 10, 2002.

Galton.org
Available online. URL: http://www.mugu.com/
galton/index.html. Posted on October 10,
2002.

Gibbs, W. J., "The Origins of Australian Meteorology" in *Federation and Meteorology*, Australian Science and Technology Heritage
Centre, 2001. Available online. URL: http://
www.austehc.unimelb.edu.au/fam/0805.html.
Posted on October 10, 2002.

History of the National Weather Service
Available online. URL: http://205.156.54.206/
pa/special/history/ Posted on October 10,
2002.

Joanne Simpson
Available online. URL: http://kopion.
uchicago.edu/drallen/remembrances/
simpson.html Posted on October 10, 2002.

Meteorology
Available online. URL: http://classics.mit.edu/
Aristotle/meteorology.html. Posted on Oc-
tober 10, 2002.

Narrative of an Explorer in Tropical South Africa
Francis Galton, 1853
Available online. URL: http://www.mugu.com/
galton/south-west-africa/travels.htm

NOAA History—A Science Odyssey
Available online. URL: http://www.history.
noaa.gov/bios/abbec.html

"Observations on two persistent degrees on a
thermometer"
Available online. URL: http://www.santes-
son.com/celsupp.htm

The Papers of Joseph Henry, Rothenberg, Marc, et
al. eds. (Columbia, S.C.: Model Editions Partnership, 2000). Electronic version based on
The Papers of Joseph Henry (Washington,
D.C.: Smithsonian Institution Press, 1996),

Vol. 7, pp. 3–104. Available online. URL:
http://adh.sc.edu

Papers of Sydney Chapman
Available online. URL: http://www.aip.org/
history/ead/alaska_chapman/19990060.html

Repeater Station
Available online. URL: http://www.connecting
thecontinent.com/ctcwebsite/ChatArchives/
chat_juliantodd.htm

Selected Classic Papers from the History of
Chemistry
Available online. URL: http://webserver.
lemoyne.edu/faculty/giunta/paperabc.html

The Catholic Encyclopedia, Volume XIV. 1912.
Robert Appleton Company.
Online Edition by Kevin Knight, 1999. Available online. URL: http://www.newadvent.
org/cathen/

The Observatory. Available online. URL: http://
www.cyberwitch.com/wychwood/Observatory/
banishingWinter.htm

The Scientific Revolution, by Richard S. Westfall-
Dsb Biographies. Robert A. Hatch-University
of Florida. Available online. URL: http://
web.clas.ufl.edu/users/rhatch/pages/03-Sci-
Rev/SCI-REV-Home/. Posted on October 10,
2002.

The Scripps Institute
Available online. URL: http://scilib.ucsd.
edu/sio/archives/siohstry/sverdrup-
biog.html

"Todd, Charles" in *Physics in Australia to 1945*, R.
W. Home, with the assistance of Paula J.
Needham, Australian Science Archives Pro-
ject, 1995. Available online. URL: http://
www.asap.unimelb.edu.au/bsparcs/physics/
P000836p.htm.

RECOMMENDED WEATHER AND CLIMATE-RELATED WEBSITES

American Meteorological Society

URL: http://www.ametsoc.org/AMS/

The American Meteorological Society promotes the development and dissemination of information and education on the atmospheric and related oceanic and hydrologic sciences. Founded in 1919, AMS publishes nine atmospheric and related oceanic and hydrologic journals, sponsors more than 12 conferences annually, and provides other programs and services.

Cloudman

URL: http://www.cloudman.com/index.htm

A great learning center for introduction to meteorology by Dr. John Day, aka "The Cloudman." A cloud atlas, articles, how to create a cloud-discovery notebook, weather links, and more. Day is the author of several popular books, including the Peterson Field Guide to the Atmosphere with Vincent Schaefer. He has produced a number of educational products for sale.

Earth Observatory

URL: http://earthobservatory.nasa.gov/

A complete site with images and data on a variety of meteorological information, articles, features and news, references, experiments you or teachers can do, and many beautiful full-color images. You can subscribe to the Earth Observatory, a free weekly newsletter about the latest stories, data, and other points of interest that have been added to the site.

Global Hydrology and Climate Center

URL: http://www.ghcc.msfc.nasa.gov/

Satellite images from around the world. The Global Hydrology and Climate Center (GHCC), established by NASA's Office of Mission to Planet Earth (now Earth Science Enterprise), is a partnership comprised of organizational elements from NASA Marshall Space Flight Center (MSFC), the Universities Space Research Association (USRA), and the Space Science and Technology Alliance of the State of Alabama (SSTA). Their charter is to build a nationally recognized program in Global Hydrology, and the primary focus of the research center is to understand the Earth's global water cycle, the distribution and variability of atmospheric water, and the impact of human activity as it relates to global climate change.

Meteorological Centers of various countries

URL: http://www.imd.ernet.in/other-met.htm

This website links to more than 65 government meteorological websites.

National Snow & Ice Data Center

URL: http://www-nsidc.colorado.edu/

You can learn anything about water, ice, glaciers, or snow. Their "State of the Cryosphere" provides an overview of the status of snow and ice as indicators of climate change. NSIDC publishes reports and a quarterly newsletter and creates and distributes data products on CD-ROM and other media. It also holds a large library collection of monographs, technical reports, and journals.

National Weather Association

URL: http://www.nwas.org/

The National Weather Association is a professional nonprofit association, incorporated in Washington, D.C., in 1975 mainly to serve individuals interested in operational meteorology and related activities. It has grown to more than 2,800 members, 50 corporate members, and more than 250 subscribers, including many colleges, universities, and weather-service agencies. International members and subscribers number more than 50.

NOAA Photo Library

URL: http://www.photolib.noaa.gov/

Hundreds of free downloadable images on a variety of weather-related subjects in a series of catalogs, including The National Severe Storms Laboratory (NSSL); the National Weather Service Historical Image Collection; The National Undersea Research Program (NURP); the NOAA Fleet Then and Now; NOAA in Space; NOAA at the Ends of the Earth; Small World; NOAA's Ark (Animals); National Estuarine Research Reserve System (NERR); NOAA Restoration Center; National Marine Fisheries Historical Image Collection; Geodesy—Measuring the Earth; Coast & Geodetic Survey Historical Image Collection; and Treasures of the Library.

Satellite Related Websites

URL: http://www.sat.dundee.ac.uk/web.html

This site contains hundreds of links to websites that have satellite or weather data. Included are links to research and education sites, software, conferences, and more.

The MET Office

URL: http://www.met-office.gov.uk/

Perhaps the oldest continually operating weather agency, the MET contains weather information for Great Britain (and the rest of the world) but also contains areas that include research and education.

The Weather Channel

URL: http://www.weather.com/

Check your local weather, download your personalized weather forecast (PC only, no Mac), or weather by email where you get a daily email weather report and even can check the weather conditions at major golf courses.

Visible Earth

URL: http://visibleearth.nasa.gov/

The purpose of NASA's Visible Earth is to provide a consistently updated, central point of access to the superset of NASA's Earth science-related images, animations, and data visualizations. These images are considered to be public domain and, as such, are freely available to the interested public-at-large, the media, scientists, and educators for reuse and/or republication.

Weather Databases

URL: http://www.internets.com/sweather.htm

This site contains links to many searchable databases relating to weather from around the world. One site has 6,000 weather links to Europe. It also has links to websites relating to weather (a recent hit list had 6,446 categories).

Weather Terminology

URL: http://www.crh.noaa.gov/lmk/terms.htm

Not quite sure what a front is? This site contains a good explanation of most weather-related terms.

WW2010 (Weather World 2010)

URL: http://ww2010.atmos.uiuc.edu/(Gh)/home.rxml

Developed by the department of atmospheric sciences (DAS) at the University of Illinois Urbana–Champaign (UIUC), The Weather World 2010 Project (WW2010) is a web framework for integrating current and archived weather data with multimedia instructional resources, using new and innovative technologies.

All of the materials have been reviewed and edited by professors and scientists from the department of atmospheric sciences (DAS) at the University of Illinois Urbana–Champaign (UIUC) and the Illinois State Water Survey.

The collection includes Online Guides, multimedia instructional modules in meteorology, and remote sensing, plus curriculum projects and classroom activities. The archives contain data and descriptions for memorable weather events. There are instructions on how to order the WW2010 Educational CD-ROM, and the Current Weather section contains current weather information for the surface, upper air, and satellite products.

INDEX

Note: Page numbers in **boldface** indicate main topics. Page numbers in *italic* refer to illustrations.